BASIC
ELECTRICITY/ELECTRONICS

Volume 1

Basic
Principles

SECOND EDITION

By Training & Retraining, Inc.

Second Edition Revised by
Robert R. Manville

Howard W. Sams & Co.
A Division of Macmillan, Inc.
4300 West 62nd Street, Indianapolis, IN 46268 USA

Preface to the Series

We live in a world of electricity and electronics. Electrical power provides us with artificial light and heat and the energy for doing many kinds of work. Electricity is the basis of radio, television, computers—the entire area of science known as electronics.

Although our advances in technology have reached the point where we can successfully break the space barrier, we are still learning new things about electricity and electronics. One of the reasons for this is that electricity has certain intangible aspects. In other words, electricity cannot be observed by our human senses in the normal manner. However, we can observe the results of the existence of electricity, and we are continually finding new ways to use it, particularly in the field of electronics.

Electronics is a relatively new science. Even though we can trace electricity back to Franklin, Bell, and Edison, electronics goes back only a few decades to discoveries and developments by such people as Marconi and De Forest. In fact, electronics didn't really become a full-fledged science until radio came into being. World War II brought about the need for rapid technological developments, and long-range radio, radionavigation, radar, sonar, etc., became realities. In the years since World War II, developments in electronics have continued at a rapid pace. Actually, the pace has been so rapid that educational and training facilities have had difficulty keeping up.

The science of electronics has expanded to such a breadth and depth that now it is really a combination of specialized technologies. Yet, these individual technologies are all based on the same fundamental principles, principles which heretofore were difficult to comprehend because of the teaching materials and methods available.

This 5-volume series represents a major step toward a unified and simplified approach to the principles of electricity and electronics. Utilizing all the modern techniques known to motivate and enhance learning, the content is designed to serve as a standard curriculum. Moreover, the programmed format has been specially prepared to provide a self-teaching tool; instructors using these volumes as classroom texts will, therefore, be able to teach the subject more objectively and with greater efficiency than ever before.

While each volume has been carefully written to "stand on its own," an understanding of the principles involved in each volume requires knowledge of the material presented in the previous volumes. The first volume in the series provides a general introduction to the overall subject of electricity and electronics. This volume is intended primarily to provide a *foundation* for the study of later volumes in the series. However, it can be used without the other volumes by the reader who requires only a relatively simple coverage of the subject.

The second volume covers basic ac and dc circuits. For the reader who has some knowledge of basic electricity, this volume can stand alone as a general text on circuit fundamentals.

The third volume is a text on the subject of tube and transistor circuits. It is written on the assumption that the reader is familiar with the principles covered in the first two volumes.

The first three volumes offer coverage of general electrical and eletronic principles. They provide the basis for further study of a general or specialized nature.

The fourth and fifth volumes deal with specialized areas of study. If the reader already has a thorough understanding of the material presented in the earlier parts of the series, either of the last two volumes can be used alone as a text in its specialized field—test equipment and servicing in Volume 4, and motors and generators in Volume 5. Volume 5 also includes an index to the 5 volumes.

Many authors, editors, and consultants have contributed

to the development of this series. It is their hope that it will serve the long-felt need for a standard text that can be used as a self-teaching guide or used in any type of training course that requires an understanding of the principles of electricity and electronics.

TRAINING & RETRAINING, INC.

Acknowledgments

Grateful acknowledgment is made to all those who participated in the preparation, compilation, and editing of this series. Without their valuable contributions this series would not have been possible.

In this regard, prime consideration is due Bernard C. Monnes, Educational Specialist, Navy Electronics School, for his excellent contributions in the areas of writing, editorial organization, and final review of the entire series. The finalization of these volumes, both as to technical content and educational value, is due principally to his tireless and conscientious efforts.

Grateful appreciation is also extended to Lt. Loren Worley, USN, and Ashley G. Skidmore, BUSHIPS, Dept. of the Navy, for their original preparatory contributions and coediting of this series. We also want to thank Irene and Don Koosis, Raymond Mungiu, George V. Novotny, and Robert J. Brite for their technical writing and contributions to the programmed method of presentation. Special thanks to Robert L. Snyder for his initial preparation and organizational work on the complete series.

Finally, special thanks are due the Publisher's editorial staff for invaluable assistance beyond the normal publisher-author relationship.

TRAINING & RETRAINING, INC.

Contents

CHAPTER 15

Introduction

This first volume in the series carefully explains the basic principles that are the foundation for understanding electricity and electronics. Following a unique method of presentation, these principles are related through simple analogies to devices with which you are familiar. You will learn that the principles are the same for both electricity and electronics and that they are not difficult to master. When you complete the volume, you will be able to understand what makes electrical and electronic devices work, and to discuss with confidence the application of these principles.

The knowledge gained from this volume will serve as an excellent foundation for further studies in the vast fields of electricity and electronics. The conventional method of learning these subjects is through study of the many individual parts, leaving it to the student to tie them together when he has finished. Experience has shown that this approach is not always successful. Therefore, this text develops only the basic principles, applies them immediately to familiar devices, and summarizes their applications in electronic equipment. In other words, this volume presents a "big picture" of the electrical/electronic field in a manner that is easily understood. The reader can study the subject without fear of becoming lost in details; he will always be able to relate what he learns to appropriate applications in the "big picture." This approach has been tried and proved successful in training thousands of students.

WHAT YOU WILL LEARN

This volume clearly explains the principles of voltage, current, and resistance, as well as their relationships to each other. You will learn the basic requirements of dc and ac circuits, how to use meters as test instruments, how parts are connected, how to read schematic diagrams, and how proper soldering techniques are carried out. You will also learn about the construction and operation of coils, capacitors, transformers, diodes, transistors, and vacuum tubes. You will discover how basic electrical devices work, including telephone systems, radio and television transmitters and receivers, etc. In addition, the text describes experiments that you can perform to improve your understanding of some of the more important principles.

WHAT YOU SHOULD KNOW BEFORE YOU START

The only prerequisites for learning electricity and electronics from this text are an ability to read and a desire to learn. All terms are carefully defined. Enough math is used to give precise interpretation to important principles, but if you know how to add, subtract, multiply, and divide, the mathematical expressions will give you no trouble. You will be shown how to interpret the meanings of simple mathematical expressions; and, incidentally, interpretation is more important to your learning than actually solving the problems.

WHY THE TEXT FORMAT WAS CHOSEN

During the past few years, new concepts of learning have been developed under the common heading of programmed instruction. Although there are arguments for and against each of the several formats or styles of programmed textbooks, the value of programmed instruction itself has been proved to be sound. Most educators now seem to agree that the style of programming should be developed to fit the teaching needs of the particular subject. To help you progress successfully through this volume, a brief explanation of the programmed format follows.

Each chapter is divided into small bits of information presented in a sequence that has proved best for learning purposes. Some of the information bits are very short—a single sentence in some cases. Others may include several paragraphs. The length of each presentation is determined by the nature of the concept being explained and the knowledge the reader has gained up to that point.

The text is designed around two-page segments. Facing pages include information on one or more concepts, complete with illustrations designed to clarify the word descriptions used. Self-testing questions are included in most of these two-page segments. Many of these questions are in the form of statements that require one or more missing words be supplied; other questions are either multiple-choice or simple essay types. Answers are given on the succeeding page, so you will have the opportunity to check the accuracy of your response and verify what you have or have not learned before proceeding. When you find that your answer to a question does not agree with that given, you should restudy the information to determine why your answer was incorrect. As you can see, this method of question-answer programming ensures that you will advance through the text as quickly as you are able to absorb what has been presented.

The beginning of each chapter features a preview of its contents, and a review of the important points is contained at the end of the chapter. The preview gives you an idea of the purpose of the chapter—what you can expect to learn. This helps to give practical meaning to the information as it is presented. The review at the completion of the chapter summarizes its content so that you can locate and restudy those areas which have escaped your full comprehension. And, just as important, the review is a definite aid to retention and recall of what you have learned.

HOW YOU SHOULD STUDY THIS TEXT

Naturally, good study habits are important. You should set aside a specific time each day to study in an area where you can concentrate without being disturbed. Select a time when you are at your mental peak, a period when you feel most alert. Here are a few pointers you will find helpful in getting the most out of this volume.

1. Read each sentence carefully and deliberately. There are no unnecessary words or phrases; each sentence presents or supports a thought which is important to your understanding of electricity and electronics.

2. When you are referred to or come to an illustration, stop at the end of the sentence you are reading and study the illustration. Make sure you have a mental picture of its general content. Then continue reading, returning to the illustration each time a detailed examination is required. The drawings were especially planned to reinforce your understanding of the subject.

3. At the bottom of most right-hand pages you will find one or more questions to be answered. Some of these contain "fill-in" blanks. In answering the questions, it is important that you actually do so in writing, either in the book or on a separate sheet of paper. The physical act of writing the answers provides greater retention than merely thinking of the answer. Writing will not become a chore since most of the required answers are short.

4. Answer all questions in a section before turning the page to check the accuracy of your responses. Refer to any of the material you have read if you need help. If you don't know the answer even after a quick review of the related text, finish answering any remaining questions. If the answers to any questions you skipped still haven't come to you, turn the page and check the answer section.

5. When you have answered a question incorrectly, return to the appropriate paragraph or page and restudy the material. Knowing the correct answer to a question is less important than understanding *why* it is correct. Each section of new material is based on previously presented information. If there is a weak link in this chain, the later material will be more difficult to understand.

6. In some instances, the text describes certain principles in terms of the results of simple experiments. The information is presented so that you will gain knowledge whether you perform the experiments or not. However, you will

gain a greater understanding of the subject if you do perform the suggested experiments.

7. Carefully study the review, "What You Have Learned," at the end of each chapter. This review will help you gauge your knowledge of the information in the chapter and actually reinforce your knowledge. When you run across statements you don't completely understand, re-read the sections relating to these statements, and re-check the questions and answers before going to the next chapter.

This volume has been carefully planned to make the learning process as easy as possible. Naturally, a certain amount of effort on your part is required if you are to obtain the maximum benefit from the book. However, if you follow the pointers just given, your efforts will be well rewarded, and you will find that your study of electricity and electronics will be a pleasant and interesting experience.

1

The World of Electricity and Electronics

what you will learn

You are about to become acquainted with the fascinating world of electricity and electronics. You are going to learn what electricity is, what it does, and how it does it. You will use this information to obtain a better understanding of what electrical and electronic devices are all about, how they work, and how to test and repair them.

WHAT IS ELECTRICITY?

Electricity is a combination of a force called **voltage** and the movement of invisible particles known as **current**.

Voltage

The force of voltage can be compared to the force of a water pump. The force of a pump moves water through a distribution system, generally an arrangement of pipes. Voltage is the force which causes electric current to flow through a system of wires.

Q1-1. Voltage is a _____.

Q1-2. The force in electricity is _____.

Q1-3. The force of voltage is something like the force of a

_____ _____.

Current

Current, the movement of invisible particles, causes electrical and electronic devices to operate. We cannot see current, but we can determine its presence by the effects it produces.

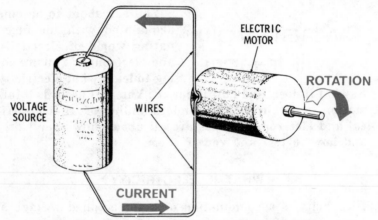

Fig. 1-1. When current flows, something happens.

Current flows through the wires of an electrical or electronic device much the same as water flows through pipes.

Voltage is the electrical force that causes current to flow.

Current consists of invisible atomic particles called **electrons.**

There is an important difference between current in wires and water in pipes, however. Water can flow out of a broken pipe, but current cannot flow out of a broken wire. In fact, current will not flow anywhere in the broken wire. When the wire is broken, the force of the voltage is removed from across the motor.

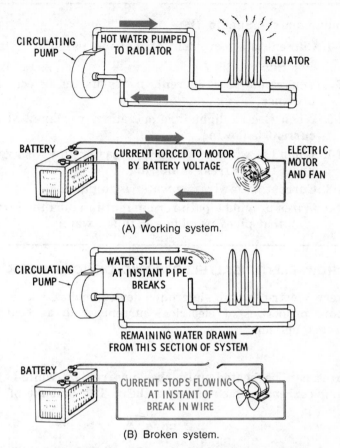

(A) Working system.

(B) Broken system.

Fig. 1-2. The flow of water and current are not exactly the same.

Q1-4. Current is a movement of invisible particles called _____.

Q1-5. You cannot see _____, but you can detect its presence.

Q1-6. When electric lights are operating, you know that _____ is flowing.

Q1-7. When the wires of an electric toaster glow red, you know that _____ is _____.

Q1-8. Current is to _____ as water is to pipes.

Q1-9. Wires provide a path for _____ in much the same way that pipes provide a path for water.

HOW ELECTRICAL/ELECTRONIC DEVICES WORK

Every electrical and electronic device makes use of one or more properties of electrical current, such as heat and electromagnetism.

Heat

Wires can be heated until they are red or white-hot by causing current to flow through them. The amount of heat

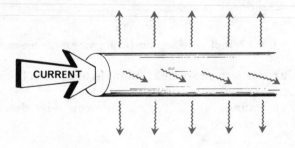

Fig. 1-3. Current flow heats wires.

given off by a wire is determined by the type of metal in the wire and the quantity of current that is forced through it. High current flow produces more heat in the same size and type of wire than low current flow. If the current is the same,

MOST ELECTRIC IRONS HAVE CONTROLS
THAT REGULATE THE AMOUNT OF HEAT
DEVELOPED BY THE HEATING ELEMENT (WIRE).

SIZE AND TYPE OF WIRE FOR
TOASTERS ARE SELECTED FOR
MOST EFFICIENT TOASTING HEAT.

Fig. 1-4. Devices that use heat produced by current flow through wires.

a smaller wire gives off more heat than one that is larger in diameter. Also, some metals produce more heat than others as the result of current flow.

In fact, manufacturers select the size and type of wire that will produce a desired amount of heat. To do this, they must know the amount of current that will flow through it.

Electrical Appliances—Electrical appliances, such as toasters, irons, heaters, and broilers, make use of the heat produced by current flowing through a wire.

Electric Lights—The filament wire in an electric light bulb is heated white-hot by the current flowing through it.

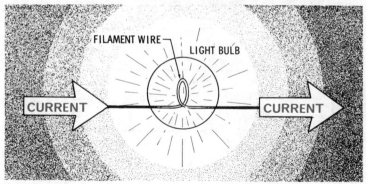

Fig. 1-5. Current flow heats a lamp filament white hot.

Q1-10. Current flow _____ wires.

Q1-11. The heat caused by _____ _____ is used to toast, iron, heat, and broil.

Q1-12. Current flow _____ the filament of an electric light bulb white-hot.

Electromagnetism

When current flows through a coil of wire, the coil acts like a magnet. This can be proved by experimenting with an electromagnet like the one shown in Fig. 1-6.

Current flowing through a wire develops a magnetic field. This field is called an **electromagnetic force** because it is the result of the flow of electric current. If, as shown in Fig. 1-6, the magnetic field passes through certain kinds of metal, such as soft iron, the metal will become **magnetized** and take on the properties of a magnet.

The electromagnet retains its magnetic capability—continues to attract iron filings—as long as current flows through the coil. When the current stops, the metal gradually loses its effectiveness as a magnet.

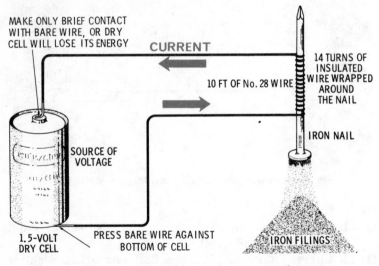

MAKE ONLY BRIEF CONTACT WITH BARE WIRE, OR DRY CELL WILL LOSE ITS ENERGY

CURRENT

14 TURNS OF INSULATED WIRE WRAPPED AROUND THE NAIL

10 FT OF No. 28 WIRE

IRON NAIL

SOURCE OF VOLTAGE

1.5-VOLT DRY CELL

PRESS BARE WIRE AGAINST BOTTOM OF CELL

IRON FILINGS

Fig. 1-6. Current flow produces magnetism.

Electric Motors—To convert electrical energy to mechanical energy electric motors make use of the magnetic forces created by current flow in a coil of wire.

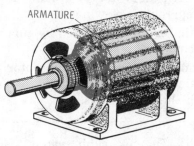

ARMATURE

Fig. 1-7. Electric motors are turned by magnetic forces.

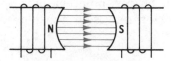

CURRENT-CARRYING WIRES COILED AROUND METAL POLE PIECES DEVELOP A MAGNETIC FIELD.

CURRENT THROUGH COILS WRAPPED AROUND THE ARMATURE (ROTATING PART) ALSO PRODUCES A MAGNETIC FIELD.

MAGNETIC FIELD FROM THE ARMATURE ENGAGES THE MAGNETIC FIELD BETWEEN THE POLE PIECES. THE PUSH AND PULL BETWEEN THE FIELDS CAUSE THE ARMATURE TO ROTATE ON ITS SHAFT.

Fig. 1-8. Magnetic fields in a motor cause rotation.

The magnetic forces in the motor are created by current flowing through the motor coils.

Q1-13. A(an) _____ _____ forms around a wire through which current is flowing.

Q1-14. _____ forces are created by current flowing in motor coils.

Q1-15. The armature of an electric motor is turned by _____ forces created by _____ flowing through the motor coils.

25

The Telegraph—The telegraph system also makes use of magnetic forces. Current flowing through a coil of wire creates magnetic forces to operate a buzzer, or other noise producers. The sounds from the buzzer represent the dots and dashes sent by the operator.

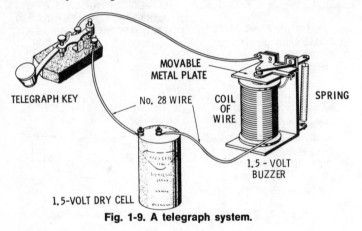

Fig. 1-9. A telegraph system.

When the telegraph key is closed (pressed down), current flows from the battery through the coil of wire. The resulting magnetic force causes a movable metal plate to be attracted to the soft-iron core of the coil, producing a buzzing sound. In this manner dots and dashes (short and long buzzes) are transmitted.

The Speaker—The speaker in a radio, television, or telephone earpiece is an example of another familiar device operated by magnetic forces created by current flowing in a coil of wire.

The speaker consists of a permanent magnet and a coil of wire cemented to a paper cone. Electrical currents which represent voice, music, or other sound flow through the coil.

Magnetic forces created by these currents cause the coil and cone to be attracted and repelled by the permanent magnet. The movement of the paper cone creates corresponding changes in air pressure heard as sounds.

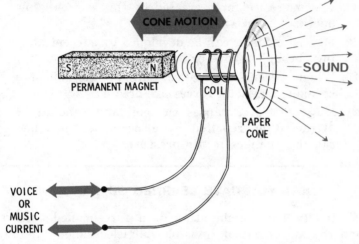

Fig. 1-10. A speaker.

Later in this volume you will learn how sounds are converted into currents that operate speakers.

Q1-16. _____ _____ in the telegraph are created by current flowing in a coil of wire.

Q1-17. Closing the key allows _____ to _____ through a coil of wire.

Q1-18. The magnetic force created by this _____ operates a buzzer.

Q1-19. Magnetic forces can be developed by current flowing through a wire or a(an) _____.

Q1-20. When currents are passed through the coil of a speaker, _____ _____ are created.

Q1-21. Magnetic force causes the coil and cone to be _____ or _____ by a permanent magnet, causing changes in air pressure.

Your Answers Should Be:

A1-16. Magnetic forces in the telegraph are created by current flowing in a coil of wire.

A1-17. Closing the key allows **current** to **flow** through a coil of wire.

A1-18. The magnetic force created by this **current flow** operates a buzzer.

A1-19. Magnetic forces can be developed by current flowing through a wire or a **coil**.

A1-20. When currents are passed through the coil of a speaker, **magnetic forces** are created.

A1-21. Magnetic force causes the coil and cone to be **attracted** or **repelled** by a permanent magnet, causing changes in air pressure.

WHAT YOU HAVE LEARNED SO FAR

1. Electricity is a combination of a force called voltage and the movement of invisible particles called current.
2. Voltage, as a force, is similar to the force developed by a water pump. Current flows through wires in much the same way that water travels through pipes.
3. All electrical or electronic devices make use of one or more of the effects produced by current flow
4. Electrical current causes wires to heat. Toasters, irons, heaters, broilers, and lights are examples of devices which use this electrical effect.
5. Current flowing through a wire or coil develops a magnetic field. Such magnetic forces are used in motors, the telegraph, and speakers.
6. A magnetic (sometimes called electromagnetic) field can be used to move a metallic piece. Speakers and motors make use of two magnetic fields that either repel or attract each other. As a result, speaker cones vibrate to develop sound and motor armatures rotate.

Current can cause other effects in addition to heating wires and producing magnetic forces. You will learn about these effects later.

HOW ELECTRICAL AND ELECTRONIC
DEVICES ARE USED

Jobs in which heat is necessary can be performed by using electric current to heat wires. Jobs calling for mechanical motion can be performed by the forces developed by magnetic fields. Transmitting telegraph messages makes use of the magnetic forces in a buzzer coil.

These simple jobs are accomplished by using simple devices. However, even complex jobs, such as sending and receiving telephone or radio messages, sending and receiving television pictures, and completely controlling manufacturing processes, are also accomplished by using simple devices.

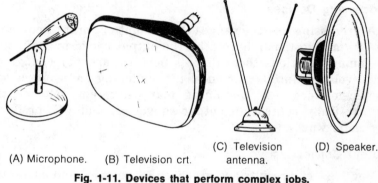

(A) Microphone. (B) Television crt. (C) Television antenna. (D) Speaker.

Fig. 1-11. Devices that perform complex jobs.

Many complex jobs are performed by combinations of simple devices. That is, two or more simple devices work together to perform a complex job. These devices fall into three categories: (1) input converters, (2) processing devices, and (3) output converters.

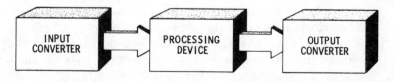

INPUT CONVERTER   PROCESSING DEVICE   OUTPUT CONVERTER

Q1-22. Simple devices that can be combined to perform complex jobs are classified as _____ _____, _____, and _____ _____.

Your Answer Should Be:

A1-22. Simple devices that can be combined to perform complex jobs are classified as **input converters, processing devices,** and **output converters.**

Input Converters

The purpose of the **input converter** in any electrical or electronic process is to convert some form of energy, such as sound, light, heat, pressure, etc., into voltage and current. These forms of energy can be processed by electrical or electronic devices after they are converted into voltage and current.

Processing Devices

A **processing device** changes the amount or form of current and voltage to that required by the output converter. Among the many functions these devices perform are (1) changing small voltages into larger ones, (2) switching and connecting telephone lines, (3) combining many different voltages to automatically control an output converter, and (4) producing radio waves.

Output Converters

Very few end results are produced by voltages and currents alone. Therefore, devices are needed which convert voltage or current into radio waves, sound, motion, heat, pictures, or other useful forms. These devices are called **output converters.**

WHAT YOU HAVE LEARNED

1. What electricity is.
2. How electricity and electronic devices operate.
3. How electrical and electronic devices are used.
4. Even the most complex jobs are accomplished by using three kinds of devices—input converters, processing devices, and output converters.
5. Any device which converts some form of energy into voltage and current performs the function of an input converter.

6. Processing devices change the input voltage and current into a form suitable for operating an output converter.
7. Output converters convert voltage and current from a processing device into some useful form.

Q1-23. The telephone mouthpiece is an input converter. It converts _____ into voltage and current.

Q1-24. A thermostat in a home heating system converts _____ into mechanical motion to open and close a switch.

Q1-25. The circuits of a radio receiver, which change electricity from radio waves (input) to the form of electricity required by the speaker (output), can be thought of as a(an) _____ _____.

Q1-26. Electric lamps convert voltage and current into _____.

Q1-27. A speaker converts voltage and current into _____.

Q1-28. Electric toaster filaments (wires) convert voltage and current into _____.

Q1-29. Television picture tubes convert voltage and current into _____.

Q1-30. Two results of current flowing through a wire are the development of _____ and _____ _____.

Q1-31. The flow of current through a wire is caused by an electrical characteristic called _____.

Q1-32. _____ and _____ are required to operate electrical and electronic devices.

Your Answer Should Be:

A1-23. The telephone mouthpiece is an input converter. It converts **sound** into voltage and current.

A1-24. A thermostat in a home heating system converts **heat** into mechanical motion to open and close a switch.

A1-25. The circuits of a radio receiver, which change electricity from radio waves (input) to the form of electricity required by the speaker (output), can be thought of as a **processing device.**

A1-26. Electric lamps convert voltage and current into **light.**

A1-27. A speaker converts voltage and current into **sound.**

A1-28. Electric toaster filaments (wires) convert voltage and current into **heat.**

A1-29. Television picture tubes convert voltage and current into **pictures** (or images).

A1-30. Two results of current flowing through a wire are the development of **heat** and **magnetic fields.**

A1-31. The flow of current through a wire is caused by an electrical characteristic called **voltage.**

A1-32. **Voltage** and **current** are required to operate electrical and electronic devices.

SAMS.

Sams books cover a wide range of technical topics. We are always interested in hearing from our readers regarding their informational needs. Please complete this questionnaire and return it to us with your suggestions. We appreciate your comments.

Book Mark Book

1. Which brand and model of computer do you use?
☐ Apple _____
☐ Commodore _____
☐ IBM _____
☐ Other (please specify) _____
_____

2. Where do you use your computer?
☐ Home   ☐ Work

3. Are you planning to buy a new computer?
☐ Yes   ☐ No
If yes, what brand are you planning to buy? _____

4. Please specify the brand/type of software, operating systems or languages you use.
☐ Word Processing _____
☐ Spreadsheets _____
☐ Data Base Management _____
☐ Integrated Software _____
☐ Operating Systems _____
☐ Computer Languages _____

5. Are you interested in any of the following electronics or technical topics?
☐ Amateur radio
☐ Antennas and propagation
☐ Artificial intelligence/
  expert systems
☐ Audio
☐ Data communications/
  telecommunications
☐ Electronic projects
☐ Instrumentation and measurements
☐ Lasers
☐ Power engineering
☐ Robotics
☐ Satellite receivers

6. Are you interested in servicing and repair of any of the following (please specify)?
☐ VCRs _____
☐ Compact disc players _____
☐ Microwave ovens _____
☐ Television _____
☐ Computers _____
☐ Automotive electronics _____
☐ Mobile telephones _____
☐ Other _____

7. How many computer or electronics books did you buy in the last year?
☐ One or two   ☐ Three or four
☐ Five or six   ☐ More than six

8. What is the average price you paid per book?
☐ Less than $10   ☐ $10-$15
☐ $16-$20   ☐ $21-$25   ☐ $26+

9. What is your occupation?
☐ Manager
☐ Engineer
☐ Technician
☐ Programmer/analyst
☐ Student
☐ Other _____

10. Please specify your educational level.
☐ High school
☐ Technical school
☐ College graduate
☐ Postgraduate

11. Are there specific books you would like to see us publish? _____
_____
_____

Comments _____
_____
_____

Name _____
Address _____
City _____
State/Zip _____

21501

Book Markram
Book Markram

BUSINESS REPLY CARD

FIRST CLASS   PERMIT NO. 1076   INDIANAPOLIS, IND.

POSTAGE WILL BE PAID BY ADDRESSEE

HOWARD W. SAMS & CO., INC.

ATTN: Public Relations Department

P.O. BOX 7092

Indianapolis, IN 46206

NO POSTAGE
NECESSARY
IF MAILED
IN THE
UNITED STATES

2

Basic Electrical Circuits

what you will learn

You are now going to learn what electrical circuits are, what they consist of, and what each device in the circuit does. You will become more familiar with voltage and current and will learn the difference between direct current (dc) and alternating current (ac). You will become acquainted with electrical diagrams and construction of circuits.

COMPLETE ELECTRICAL CIRCUITS

A circuit is usually defined as any path that returns to its starting point. In electricity, current makes a complete trip through an **electrical circuit**.

If the circuit is not complete, current does not flow. Current flows only if the path through the circuit is complete. A broken wire, a loose connector, or a switch in the OFF position will prevent current from flowing.

You have now learned two important facts regarding the flow of current. A **voltage source causes** current to flow, and a **complete circuit allows** current to flow.

Q2-1. In electricity, a circuit provides a path for _____ to make a _____ trip.

Q2-12. Current flow is caused by a(an) _____ _____ and permitted by a(an) _____ _____.

HOW ELECTRICAL CIRCUITS ARE MADE

All electrical circuits consist of the basic units shown in Fig. 2-1. The device being operated, of course, may be any electrical or electronic device. In fact, many electrical circuits contain more than one device to be operated.

Now that you are familiar with the basic units of an electrical circuit, you are ready to learn more about each part.

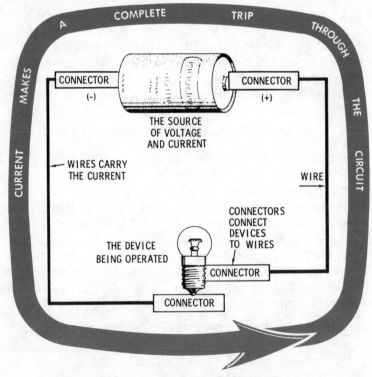

Fig. 2-1. A basic electrical circuit.

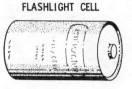

FLASHLIGHT CELL

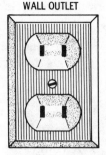

WALL OUTLET

Fig. 2-2. Familiar sources of voltage and current.

Voltage Source

A battery is an example of a voltage source. As you recall, a voltage source is also a source of current.

The electrical wall socket (or outlet) is another widely used source of voltage and current. The outlet is part of another circuit that has a generator as a voltage source. There may be many miles of wire between the generator and the outlet.

The battery shown in Fig. 2-2 is more properly called a cell. A cell was originally considered to be a storage device. Cells, such as those used in a flashlight, develop 1.5 volts each. Some new rechargeable cells only develop 1.2 volts each. Technically, a battery is a device constructed of two or more cells. However, through long usage, a cell is often called a battery.

Q2-3. List the four basic units of an electrical circuit.

Q2-4. A lamp will light only when it is part of a(an) _____ _____.

Q2-5. A(an) _____ joins wires to a voltage source or operating device to make a complete current path.

Q2-6. A(an) _____ _____, such as a battery, is also a source of current.

Q2-7. Voltage sources of the type used in a flashlight are more properly called _____.

Q2-8. A single flashlight cell is a source of _____ volts.

Q2-9. If a 6-volt battery has four 1.5-volt cells, a 12-volt battery will have _____ 1.5-volt cells.

Your Answers Should Be:

A2-3. The four basic units of an electrical circuit are:
1. Voltage source.
2. Device being operated.
3. Wires.
4. Connectors (terminals).

A2-4. A lamp will light only when it is part of a **complete circuit.**

A2-5. A **connector** joins wires to a voltage source or operating device to make a complete current path.

A2-6. A **voltage source,** such as a battery, is also a source of current.

A2-7. Voltage sources of the type used in a flashlight are more properly called **cells.**

A2-8. A single flashlight cell is a source of **1.5** volts.

A2-9. If a 6-volt battery has four 1.5-volt cells, a 12-volt battery will have **eight** 1.5-volt cells. (If a 12-volt battery provides twice as much voltage as a 6-volt battery then it must have twice as many cells.)

Voltage Source Connections—All sources of voltage (and current) have at least two connections.

The source of voltage and current in an electrical circuit is similar to a pump in a water system. The **pump** provides both the **pressure** and the **water** to cause a flow through the

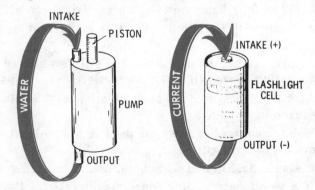

Fig. 2-3. Electricity is like water in many ways.

water system. A **voltage source** provides **electrical pressure** (voltage) and **current** (electrical equivalent of the water) to cause a flow through an **electrical circuit.**

Like the water pump, **the source of voltage and current requires an input connection and an output connection.**

WARNING: Caution must always be observed when working near voltage sources or circuits. If you come in contact with both connections of the source, **your body becomes a circuit,** and current will flow through you. This can cause painful burns and even death. If you touch only one side of the source or a single wire leading to it, **be sure you do not touch a pipe or other metal surface in contact with the ground.** This precaution is necessary because many voltage sources have one connection wired to ground.

Conductors and Insulators

Wires provide a path for electric current just as pipes provide a path for water. Metals such as copper and aluminum are most commonly used in the manufacture of electrical wire. Their atomic structures make these metals good **conductors** of current. Silver is the best conductor but is much more expensive than other metals. Other more economical metals, such as copper and aluminum, are good conductors and are quite easily formed into wire. When connected into a circuit, wire is referred to as a **conductor.**

Most nonmetals are very poor conductors of electric current. These materials are called **insulators.** Rubber and plastic are two commonly used materials for insulators because they are flexible, easily molded, and can be readily cut when necessary. Because of their better insulating qualities, glass and ceramic materials are used where high-voltage insulators are required.

Q2-10. How many connections must be made to a voltage source?

Q2-11. Materials which provide an easy path for current are called _____. Those which do not provide an easy path are called _____.

Working With Wire

Practically all wire used in electrical and electronic work consists of a conducting metal (usually circular in cross section) covered with insulation. The insulation prevents **undesired connections** to and between conductors.

Exceptions include wire used for heating purposes. In these cases, the heating element (wire) is wrapped or formed on an insulating material or supported in air (a nonconductor) between insulators.

If a bare wire comes in contact with another conductor or other metal in an electrical unit, a **short circuit** develops. Current will flow through the **short** instead of the complete circuit containing the operating device. For this reason, wire should be handled with sufficient care to ensure that its insulation is not damaged.

Wire Stripping—In order to join a wire to a connector, a length of insulation must be removed from the wire. A metal-to-metal connection is required to permit current flow.

The process of removing insulation is called **wire stripping** and is properly accomplished by a tool called a **wire stripper**. With care both wire cutting and stripping can be done with **diagonal cutters**. See Fig. 2-4.

When stripping wire with a wire stripper, place the wire in the correct size (marked on stripper), squeeze the handles, and pull the insulation off the wire with the stripper.

Do not squeeze the plier handles too tightly when attempting to remove the insulation from the wire. Just break the surface of the insulation with the cutting head. The cut need not go through to the wire. A steady pull should then part or tear the remaining insulation. Placing the index finger between the handles prevents the cutters from closing completely and nicking or cutting the wire.

NARROW
HEAD

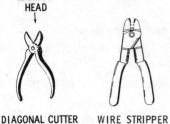

DIAGONAL CUTTER   WIRE STRIPPER

TO OPEN JAWS OF THE CUTTER:
Spread the handles using the index finger

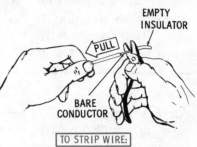

EMPTY
INSULATOR

PULL

BARE
CONDUCTOR

TO CUT WIRE:
Place the wire between the jaws
of the cutter and squeeze the
handles together

TO STRIP WIRE:
1. Place the wire between the jaws
   (near the pivot).
2. Squeeze the handles just enough
   to cut the insulation.
3. Pull the long end of the wire away from
   the cutter. (Firmly grip the insulation
   with the jaws of the cutter).

Fig. 2-4. Cutting and stripping wire.

Q2-12. Nearly all metals will conduct _____.

Q2-13. Copper or aluminum are used in electric wires
because they are good _____.

Q2-14. Materials that are nonconductors of current are
called _____.

Q2-15. Insulation is used on wires to (make, prevent)
contact with other conductors of current.

Q2-16. Undesired contact between two conductors is
called a(an) _____ _____.

Q2-17. _____ _____ is the process of removing insulation from a wire. It can be accomplished by _____
_____ or _____ _____.

Devices

Current from a voltage source operates devices such as electric light bulbs, heaters, and motors. Radio and television receivers are also operated by current from voltage sources. These devices process voltage and current contained in received radio waves by changing the input energy into sound and pictures. The voltage sources make it possible for these devices to perform this process.

Connections—As stated earlier, all electrical devices must have **two or more** connections to a circuit. These connections are used to join conductors to the device, thus completing the circuit and permitting current to flow into and out of the device.

Operation—The voltage source operates the device by forcing current through the circuit. All connections must be made in the circuit, including those at the device and the source. Current will then be able to flow through the device and cause it to operate.

Connectors

The terms **connectors** and **terminals** are often used interchangeably. A **connector,** however, is normally thought of as being a mechanical part, such as a battery clamp, used to connect a conductor to a device. A **terminal,** on the other hand,

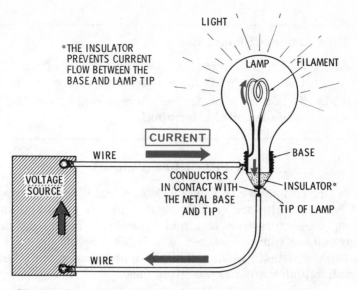

LIGHT

LAMP

FILAMENT

*THE INSULATOR
PREVENTS CURRENT
FLOW BETWEEN THE
BASE AND LAMP TIP

CURRENT

WIRE

VOLTAGE
SOURCE

CONDUCTORS
IN CONTACT WITH
THE METAL BASE
AND TIP

BASE

INSULATOR*

TIP OF LAMP

WIRE

Fig. 2-5. Connection and operation of an electrical device.

is a point on a device where a connection can be made—a
screw or other contact point.

The illustration in Fig. 2-5 shows how connections are made
to a voltage source and an operating device. The lamp will
light with the bare conductors merely touching the lamp terminals. In practice, however, the lamp is placed in a socket
and the wires connected to the socket terminals.

Wires, connectors, and terminals allow current to flow in
a circuit because they are made of conducting metals. Care
must be taken, however, when joining these parts to each
other. Metal at the contact points must be clean and free from
the insulating properties of dirt, grease, etc. Sandpaper can
be used to clean these junction points when necessary. After
a wire has been stripped, it should be cleaned of any remaining
insulation.

When connecting a wire to a terminal, make sure the screw
or clamp makes a tight connection. For current to flow, all
parts of the circuit must be connected.

Q2-18. To permit current to flow into and out of a device,
the device must have at least _____ connections.

Q2-19. That part of a device where a connection can be
made is called a _____.

A PRACTICAL CIRCUIT

The circuit shown in Fig. 2-6 demonstrates the way in which all basic circuits are connected. It contains a voltage source, wire, connectors (or terminals), and an operating device. The voltage source pictured is a large 1.5-volt dry cell used in some doorbell systems. This is a practical circuit because it will actually work and is often used.

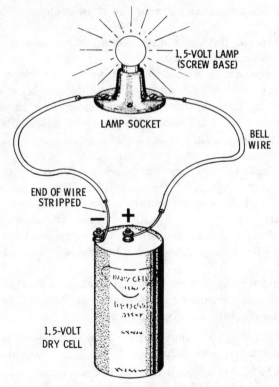

Fig. 2-6. A practical circuit.

Open Circuits

If all the connections are made as shown in the illustration, the lamp will light. If any of the connections are not properly made, the lamp will not light—a condition known as an **open circuit**. An open circuit represents a condition that prevents the flow of current. In other words, the circuit is not complete.

Closed Circuits

A **closed circuit** has all of its connections made and forms a complete path through which current can flow.

SWITCHES

Since it is often desirable to open and close a circuit, nearly all circuits contain some form of **switch**.

Knife Switch

One simple type of switch is called a **knife switch**. It was given this name because it has an element resembling the blade of a knife.

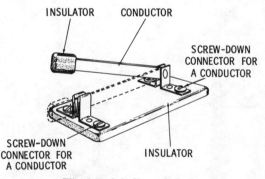

Fig. 2-7. A knife switch.

Q2-20. The lamp in the illustration on the opposite page lights because it is a(an) (open, closed) circuit.

Q2-21. Disconnecting one of the wires will develop a(an) (open, closed) circuit.

Q2-22. To permit opening and closing a circuit, a(an) _____ can be connected into it.

Q2-23. A (an) _____ _____ is a basic type of switch.

The basic circuit just explained can be reconnected to include a knife switch. The illustration in Fig. 2-8 shows how the connections are made. Be sure you understand what happens to the flow of current when the switch is open (position shown) and when it is closed.

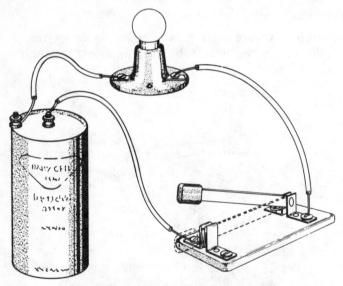

Fig. 2-8. A practical circuit with switch.

There are many other types of switches, some of which you have used. For example, there are switches on the walls of your home, on the front of your appliances, and on the dashboard of your car. Many operate on the knife-switch principle. In the closed position, a metal blade makes an electrical contact between at least two conductors.

WHAT YOU HAVE LEARNED ABOUT
ELECTRICAL CIRCUITS

1. An electrical circuit provides a complete path for current flow.
2. Every electrical circuit consists of (1) a source of voltage which causes current to flow; (2) conductors which provide a path for the current; (3) electrical devices which are operated by the current; (4) connectors (terminals) to join conductors to a source or a device.
3. Voltage sources and electrical devices always have at least two connections. All connections must be made in order for current to flow through them.
4. Most metals can conduct current and are called conductors. Most nonmetals provide a very poor path for current and are called insulators.
5. A wire consists of a conductor (usually copper) covered by insulation (usually rubber or plastic). Insulation may be stripped with wire strippers or diagonal cutters.
6. An open circuit is a condition in which the current path is interrupted. A closed circuit is the same as a complete circuit. A short circuit occurs when a conductor makes an undesirable contact with another conductor or metal part.
7. Switches are designed to open and close circuits. By operating the switch, a device may be turned on or off.

REVIEW QUESTIONS (Mark them true or false.)

Q2-24. A voltage source causes current to flow if a complete circuit is provided.

Q2-25. Current will flow if there is a complete electrical path from the voltage source to the device, through the device, and back to the voltage source.

Q2-26. If otherwise complete, current will not flow in a circuit if its switch is closed.

Q2-27. When a wall switch is flipped to the ON position, the switch is open, permitting the lamp to light.

Q2-28. If the insulation on a wire is broken or damaged, it may cause a short circuit.

Your Answers Should Be:

A2-24. A voltage source causes current to flow if a complete circuit is provided. **True.**

A2-25. Current will flow if there is a complete electrical path from the voltage source to the device, through the device, and back to the voltage source. **True.**

A2-26. If otherwise complete, current will not flow in a circuit if its switch is closed. **False.**

A2-27. When a wall switch is flipped to the ON position, the switch is open, permitting the lamp to light. **False.**

A2-28. If the insulation on a wire is broken or damaged, it may cause a short circuit. **True.**

VOLTAGE AND CURRENT

In this section you will become acquainted with voltage and current measurement units. You will also become familiar with the commonly used values of these units.

Voltage

Voltage is measured in terms of a unit called a volt. A measurement unit indicates quantity or amount, as in liters of water or kilograms of sugar. As a similar unit, **volts** expresses a quantity contained in a voltage source. Although voltage is not visible like water and sugar, the number of volts expresses the amount of electrical pressure available from the source. As you remember, it is this pressure that causes current to flow. The greater the pressure (number of volts), the greater the current will be.

Current

Current is measured in terms of a unit called an **ampere**. The number of amperes (somtimes called amps) defines the amount of current that is flowing in a circuit. Some flashlight lamps (bulbs), for example, draw 0.25 **ampere** (abreviated as 0.25 A) from the voltage source.

A 100-watt lamp draws approximately 1 A from the 117-volt home electrical system. Ten amperes flow through some electric irons, toasters, and heaters. A car battery supplies 100 A or more to a starter motor.

Large and Small Values

Values of voltage and current can be very large or very small. Since it is awkward to talk and write about 500,000 volts or 0.003 A, units which are more easily handled have been developed. With this system the quantities mentioned become 500 kilovolts (kV) and 3 milliamps (mA), respectively. A **kilovolt** represents 1000 volts and a **milliamp,** 0.001 A.

Table 2-1 will help you convert from one unit to another.

Table 2-1. Conversion Table

When You See	Do This To Convert	Example
mega or M	Multiply by 1,000,000	2 megavolts is 2,000,000 volts
kilo or k	Multiply by 1000	5 kiloamperes is 5000 amps
milli or m	Divide by 1000	7 millivolts is 0.007 volt
micro or μ	Divide by 1,000,000	9 μamperes is 0.000009 ampere
nano or n	Divide by 1,000,000,000	5 nanovolts is 0.000000005 volt
pico or p	Divide by 1,000,000,000,000	5 picoamperes is 0.000000000005 ampere

Q2-29. Voltage is measured in units called _____.

Q2-30. The number of volts indicates the quantity of _____ _____ contained in a voltage source.

Q2-31. An ampere is a unit that indicates the quantity of _____.

Q2-32. Assuming that the voltage source can provide the current, what determines the number of amperes that will flow in a circuit?

Q2-33. A kilovolt is (larger, smaller) than 100 millivolts.

Q2-34. How much of an ampere is 15 microamperes?

Q2-35. Convert 16 megavolts to volts.

DIRECT CURRENT

A current that flows in one direction only is called a **direct current.** Dry cells and batteries are sources of direct current. Some types of electric generators also supply direct current. Later you will learn about a power supply which provides direct current for use within radio and tv receivers.

Is There a Direct Voltage?

Yes. A voltage which provides direct current is considered to be a direct voltage. Since direct current is abbreviated dc, the abbreviation is used to identify direct voltage as dc voltage. Direct current is often shortened to dc current, or merely dc.

Direction of Current Flow

Marking the terminals of a voltage source with plus (+) and minus (−) signs indicates the direction in which current flows in a circuit. There are two systems describing the direction of current flow—**conventional** and **electron.**

The conventional current theory was the first to be developed. Benjamin Franklin is considered to be its originator, and it is still being used in many electrical engineering texts. **Conventional current** is said to flow **from the positive** (+) voltage terminal, through the circuit, and return **to the negative** (−) voltage terminal.

The **electron current** theory, of more recent origin, permits a clearer explanation of how current flows through electronic circuits. For this reason, the electron current direction of flow will be used in this text. This theory states that current **leaves the negative** (−) terminal, flows through the circuit, and **returns to the positive** (+) terminal of the voltage source.

If you learn the rules of electron flow, conventional flow should not be confusing. You will find it easy to mentally reverse directions.

Current flow does all the work involved in the operation of any electrical or electronic device, whether it is a simple lamp or a complicated electronic computer. In any application a continuous path must be provided between the two terminals of a voltage source before current can flow.

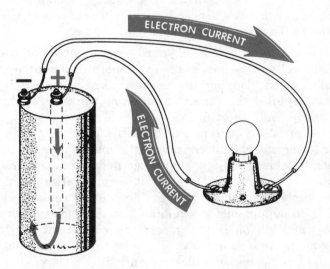

Fig. 2-9. The direction of dc electron current flow.

Q2-36. The connecting posts on the cell in the illustration are marked (+) and (−). The (+) post is the _____ terminal.

Q2-37. Inside the cell, electron current flows from the _____ terminal to the _____.

ALTERNATING CURRENT

A current that reverses its direction of flow at regular intervals is called **alternating current (ac)**. You might ask, "Why should we have a current that is constantly changing its direction?" Alternating current has certain features that make it desirable. The two main reasons are

Reason 1. Wall outlets in your home supply an ac voltage. This voltage is produced by generators located many miles away. During the earliest days of electricity, dc was supplied to homes. However, dc can be sent through lines for only short distances.

Alternating current can be easily changed to a higher or lower value. This characteristic makes possible its economical transmission over long distances—hundreds of miles in some cases. As a result, ac generating plants can be located at remote sources of water power and still be able to supply customers miles away. A good example of this application is the generating equipment at Hoover Dam in Arizona supplying power to cities on the West Coast, hundreds of miles distant.

Reason 2. The preceding chapter described input converters which convert other forms of energy into voltage and current. Many of these forms, such as sound and radio waves, occur in alternating cycles. Sound waves, for instance, are alternating areas of maximum and minimum air pressure. When converted into electricity, as in the telephone, the resulting current is also alternating; thus, the sound is faithfully transmitted.

WHAT YOU HAVE LEARNED

1. The measurement unit of electrical pressure is the volt. It defines the amount of electrical pressure available in a voltage source.
2. The measurement unit for current is the ampere, abbreviated A. Assuming that a sufficient amount of current can be supplied by the voltage source, the number of amperes that flow in a circuit is determined by the needs of the operating device. Operating devices are designed for a specified number of volts, and are so constructed as to draw the required number of amperes when operated at that voltage.
3. Volt and ampere quantities are often expressed in very large and very small numbers. To ease the task of writing or speaking of very large or very small numbers, prefixes, such as mega-, kilo-, milli-, and micro-, have been added to the basic units of volts and amperes.
4. A current that always flows in the same direction is direct current. Its abbreviation is dc, which can be used to specify dc current or dc voltage.
5. Current flows from the negative terminal of a voltage source, through the circuit, and returns to the positive terminal. Inside the voltage source, current flows from the positive to the negative terminal. This is in accordance with the electron current theory.
6. A current that reverses its direction of flow at regular intervals is called alternating current (ac). An ac voltage and current can be transmitted over long distances economically, but dc cannot. Alternating current is also the only means of converting certain types of energy into useful electrical representations.

Q2-38. A volt is a measurement of _____, and an ampere is a measurement of _____.

Q2-39. According to the electron current theory, current flows from the _____ voltage terminal, through the circuit and returns to the _____ terminal.

Your Answers Should Be:

A2-38. A volt is a measurement of **voltage,** and an ampere is a measurement of **current.**

A2-39. According to the electron current theory, current flows from the **negative** (−) voltage terminal, through the circuit, and returns to the **positive** (+) terminal.

3

How To Use Meters

what you will learn

Since volts and amperes are units of measurement, some device must be used to measure them. Devices used for this purpose are called meters. You are now going to learn about the different types of meters and how to use them to measure voltage and current. The precautions to take when handling these instruments are also discussed.

HOW DO METERS WORK?

Meters, like motors, convert electrical energy (current) into mechanical motion. In a motor, current-generated magnetic fields cause the armature to rotate. In most meters, similar magnetic fields cause a pointer to move across a scale. The position of the pointer (sometimes called a needle or indicator) when it comes to rest on the scale indicates the amount of current flowing through the meter.

Most homes and cars have meters similar in principle to those that will be discussed. An electrical meter measures consumption of house current. The gasoline, temperature, and other automobile gauges are all basically meters measuring current flow. The quantities being measured are converted to equivalent values of current.

Q3-1. Voltage and current are measured by _____.

Q3-2. The reading of a meter is taken where a pointer comes to rest on a _____.

Q3-3. Pointer movement is caused by _____ _____.

Your Answers Should Be:

A3-1. Voltage and current are measured by meters.

A3-2. The reading of a meter is taken where a pointer comes to rest on a **scale**.

A3-3. Pointer movement is caused by **magnetic fields**.

READING METERS

The illustration in Fig. 3-1 shows how a current-measuring meter is connected to measure the amount of current flowing in the circuit.

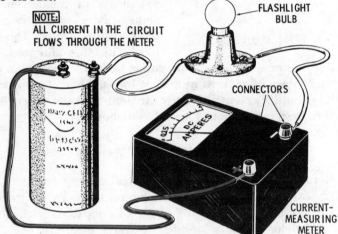

Fig. 3-1. Measuring the current drawn by a lamp.

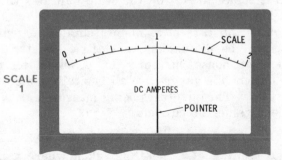

Meters are read by noting to which number (or division mark between numbers) on the scale the needle is pointing.

If the needle points to a division mark between two numbers, the decimal value of the division is added to the lower number.

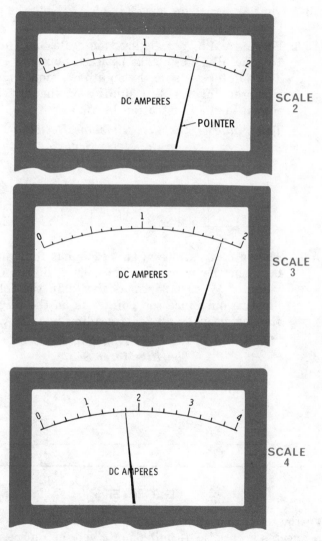

Q3-4. What is the reading for scale 1?

Q3-5. What is the value read on scale 2?

Q3-6. Scale 3 indicates a reading of _____ A.

Q3-7. What does scale 4 read?

Your Answers Should Be:

A3-4. The reading for scale 1 is **1 ampere (A).**

A3-5. The value read on scale 2 is **1.5 A.** (Note the pointer is halfway between 1 and 2 on the scale.)

A3-6. There are **1.8 A** registered on scale 3. (There are 10 equal division marks between numbers 1 and 2. The pointer rests on the eighth division, indicating a current of 1.8 A. Counting of the divisions is shown in the illustration below.)

The Value of the Ten Divisions Between 1 and 2
on the Meter Scale

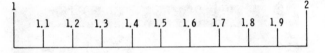

A3-7. Scale 4 reads **1.75 A.** (This scale has four divisions between the numbers. Thus, each division has a value of ¼, or 0.25 A, as shown in the following illustration. Since the pointer is on the third division between 1 and 2, its reading is 1.75 A.)

The Value of the Four Divisions Between 1 and 2
on the Meter Scale

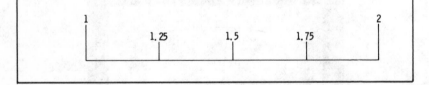

VOLTMETERS

Voltmeters are used to measure voltage. When the voltmeter is connected across the terminals of a voltage source, a current proportional to the source voltage flows through the meter mechanism. The meter scale is graduated (drawn) to give a reading in volts. The procedure for reading a voltmeter scale is similar to the current scales you have just read.

Precautions

There are two basic types of voltmeters—one for measuring dc voltage and the other for ac voltage. Be sure to use the correct one for the type of voltage to be measured. When an ac voltmeter is applied to a dc source, an incorrect measurement will occur. But when a dc meter is used to measure ac voltage, **the meter may be damaged.**

Reading a Voltmeter

As shown in the following illustration, a voltmeter scale is similar to a current-measuring scale. A value between numbers is read in the same manner as a current reading.

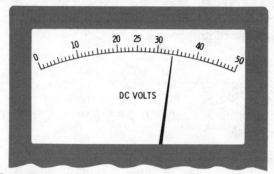

Voltage Ranges

Voltmeters are designed to read to certain maximum values. From zero to a maximum voltage is called the **range** of a voltmeter. Some commonly used ranges are 0-10 volts, 0-50 volts, 0-250 volts, and 0-1000 volts.

Always be sure that any voltage to be measured is within the range of the voltmeter you are using. A meter will be damaged if used to measure a voltage greater than the maximum value for which it is designed. Excess voltage will cause excess current to flow. As a result, the pointer may be bent in trying to move beyond the end of the scale, or meter circuits may overheat and damage delicate parts.

Q3-8. What type of meter is used to measure dc voltage?

Q3-9. How many volts are indicated in the above illustration?

Q3-10. What may happen if a voltmeter is used to measure voltages beyond its range?

AMMETERS

A current-reading meter is called an **ammeter**. It can **only** be used to measure amperes.

Current Ranges

Commonly used current ranges for work on electrical appliances are 0-10 A and 0-30 A. When working with electronic devices, ranges such as 0-500 μA, 0-10 mA, and 0-250 mA may be required.

Precautions

Ammeter precautions are the same as for voltmeters:
1. Never use dc meters for ac, or ac meters for dc.
2. Do not measure a current value that is beyond the range of the meter.

The first rule can be observed if you know the type of voltage source supplying the current. For example, you know that batteries supply dc current (and voltage), and most wall outlets supply ac current (and voltage).

The second rule can be followed as you gain experience. If your meter has a selection of ranges, always use the highest range first. Then switch to the appropriate range to obtain the most accurate reading. Quickly remove the meter leads if the pointer swings beyond the limits of the scale.

A third rule must be added to the above. Never use an ammeter to measure voltage nor a voltmeter to measure current. Each meter is designed to measure only certain electrical values. If either type of meter is used for measuring other values, it may be damaged.

MULTIMETERS

A **multimeter** is a combination voltmeter and ammeter. It can be used to measure either ac or dc voltages and currents. A multimeter is also called a **volt-ohm-milliammeter** (vom) or a **circuit analyzer**.

Reading Multimeters

A multimeter face has a combination of scales that may include several ranges of voltage and current readings. A typical multimeter scale having three ranges is shown below.

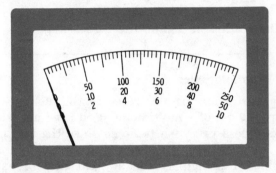

By proper front-panel settings a multimeter can be used to measure ac and dc current and voltage.

Q3-11. To measure current, use a(an) _____.

Q3-12. State three precautions that must be observed when using ammeters or voltmeters.

Q3-13. Shown here is a portion of the scale illustrated above. What is the reading on the 0-10 range?

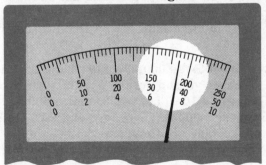

Q3-14. What is the reading on the 0-250 range?

Multimeter Characteristics

Several experiments are described in this volume. Most of them require the use of a multimeter. You need not work these experiments unless you wish to do so since the text describes the results of each one. However, you can obtain a better understanding of principles and a great deal of experience working with electrical parts and tools by performing the experiments.

Although you may not desire to purchase a multimeter until a later date, you should have some knowledge of what to look for. A good multimeter can be purchased in most electronic parts stores at a reasonable price. Or it can be ordered from one of the many mail order companies.

A multimeter from which you can obtain suitable accuracy and which has useful ranges should have the following characteristics. Each characteristic is explained in detail in Volume 4.

Sensitivity: 5000 to 10,000 ohms/volt on ac and 20,000 ohms/volt on dc.
Voltage Ranges: 0-10, 0-50, 0-250, and 0-500.
Current Ranges: 0-500 μA, 0-10 mA, and 0-250 mA.
Current and Voltage: Both ac and dc.
Resistance Ranges: 0-10 kΩ, 0-100 kΩ, and 0-1 MΩ.

A typical multimeter is shown in Fig. 3-2. Study the drawing to become familiar with the location and names of the various parts, controls, and scales. The next few pages describe each in detail.

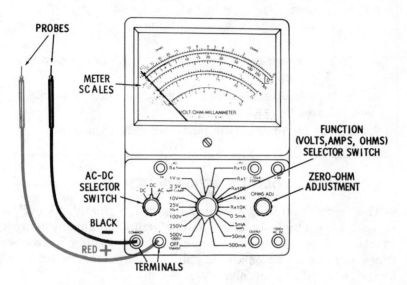

Fig. 3-2. A typical multimeter.

The front panels of some multimeters do not look like this one. Each, however, has a similar means of accomplishing the same measuring tasks.

Q3-15. Most of the experiments require the use of a
_____ _____.

Q3-16. A multimeter having an ac sensitivity of 10,000 ohms/volt will have (suitable, unsuitable) accuracy for most purposes.

Q3-17. If the ac-dc Selector Switch in Fig. 3-2 were set on dc, you would read the position of the pointer on the (first, second, third, fourth, fifth, sixth) scale.

Q3-18. What is the meaning of the "10V" marking on the Function Selector Switch? Make a guess.

Q3-19. What are the tip ends of the two test leads connected to the meter called.

VOLTAGE MEASUREMENTS

The term "multimeter" means literally "many meter." It is, in fact, a single instrument performing many measuring functions. The multimeter shown on the preceding page measures ac volts, ac amperes, dc volts, dc amperes, decibels, and ohms (to be discussed shortly).

Learning to use a multimeter well requires you to think only of the particular function for which you are using the instrument. If you are measuring dc voltage, think dc voltmeter. If the next measurement is ac amperes, change your thinking to an ac ammeter. By concentrating in this manner, you are more certain to make the proper settings and observe the appropriate measuring precautions. For this reason, the multimeter will be discussed in terms of its separate measuring functions.

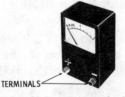

TERMINALS

Fig. 3-3. Voltmeter terminals.

Terminals—The voltmeter, like other electrical devices, has two terminals. Both terminals are connected into a circuit when using the instrument. The terminals are sometimes colored red (+) and black (−) to identify the positive (+) and negative (−) connections.

Test Leads—A voltmeter requires a pair of **test leads** to connect the meter to the circuit being tested. Test leads are lengths of flexible insulated wire. One end has a means of joining the lead to the voltmeter. The other end has a metal probe encased in an insulated handle.

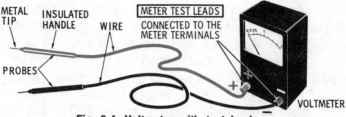

Fig. 3-4. Voltmeter with test leads.

Connections—When measuring voltage, the probes are touched to the terminals of the voltage source or device. A voltage measurement is **always** taken **across** the terminals and is **never** made **between** a terminal and an open wire.

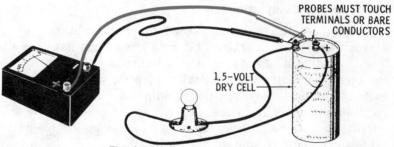

Fig. 3-5. Voltmeter connections.

Q3-20. List the six electrical quantities that a typical multimeter will measure.

Q3-21. How is ac and dc current different?

Q3-22. When measuring battery voltage, how should you think of a multimeter?

Q3-23. How are the positive and negative terminals of some voltmeters identified?

Q3-24. What part of a test lead is placed in contact with the circuit being tested?

Q3-25. A voltage is always measured _____ the terminals of a source or a device.

DC Voltmeter Connections

As you recall from the preceding chapter, a dc voltage source has both a negative and a positive terminal. The distinction between negative and positive voltage is identified by the term "polarity." The **polarity** of a dc voltage source (a battery, for example) is usually indicated in some way at its terminals. One is negative and the other positive. In a dc circuit, the terminal polarity of an operating device is the same as the supply source.

The terminals of a dc voltmeter are either colored or marked to indicate the polarity. A red color or a plus (+) mark identifies a positive terminal. Black or minus (−) indicates a negative terminal. The negative terminal of a dc voltmeter is connected through a test lead to the negative terminal (source or device) of the circuit. The other test lead is connected to the corresponding positive terminal of the meter and of the circuit.

Always observe this rule: **The polarity marking of the dc voltmeter terminal must be the same as the polarity of the voltage being measured.**

If you disobey the rule, the scale pointer will move opposite to its normal direction and may be damaged.

The illustration in Fig. 3-6 shows the proper connections to be made when measuring dc voltage.

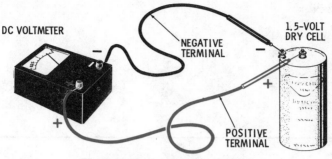

Fig. 3-6. Measuring dc voltage.

Voltage Measurements With a Multimeter

A multimeter can be adjusted by means of selector switches to measure either ac or dc voltage.

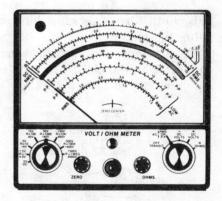

Fig. 3-7. Multimeter settings
to measure 30 volts ac.

Q3-26. Mark this figure to show the switch settings required to measure 6 volts dc.

Q3-27. Show the settings for measuring the voltage of a wall outlet.

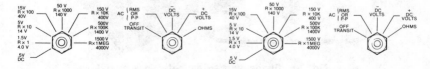

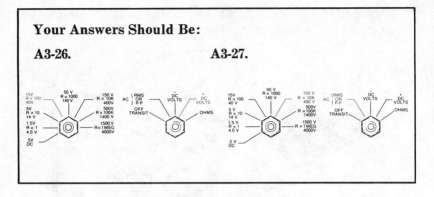
CURRENT MEASUREMENTS

Methods used to measure current with an ammeter or multimeter are different from those used to measure voltage.

Ammeter Connections

Terminals—An ammeter, like a voltmeter, has two terminals. Both terminals must be connected into the circuit when using the meter.

Connections—To measure current, the ammeter must be connected in the circuit in such a way as to allow the current being measured to flow through the meter.

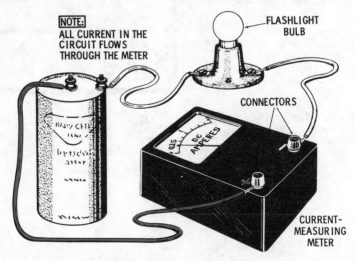

Fig. 3-8. Connecting an ammeter in a circuit.

Current Measurements With a Multimeter

Connections—Multimeter connections for measuring current are made as if the instrument were an ammeter. The circuit must be opened (usually at a terminal) and the probes inserted, one on either side of the break.

When measuring dc current, a polarity rule must be observed: **dc current should enter the negative terminal of a dc ammeter and leave by its positive terminal.** Since you know that dc current flows through a circuit from the negative to the positive terminals of a voltage source, current direction can easily be determined.

If you plan to do many experiments that require measuring currents, the board shown in Fig. 3-9 should be worth constructing. The ammeter probes are inserted into the Fahnestock clips. An experimenter's board is also available.

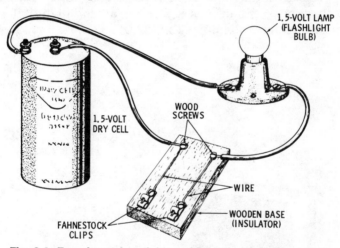

Fig. 3-9. Experiment board for making current measurements.

Settings—When a multimeter is used as an ammeter, the function switch is set to the appropriate range. In addition, the ac-dc switch is set for the kind of current (ac or dc) to be measured.

Q3-28. What is the difference between connecting a voltmeter and an ammeter into a circuit?

Q3-29. Dc current should enter the _____ terminal of a dc ammeter.

DIGITAL MULTIMETERS

Instead of an analog needle and scale, a digital multimeter (dmm) displays a value in numbers.

The basic circuits change the input signal to a dc voltage and then applies the dc voltage to an analog-to-digital converter that drives a display (Fig. 3-10).

Fig. 3-10. A 3-digit digital multimeter.

Fig. 3-11. A 3½-digit
digital multimeter.

The description of the instrument includes such terms as 3½ digits, automatic polarity, 100% overrange, and possible autoranging.

The 3½-digits (Fig. 3-11) means that the most-significant digit (digit to the left) will read only 1 or 0—that is the ½-digit. The other digits of the display will be 0, 1, 2, 3, . . . 9.

Some digital multimeters have a built-in device that enables the meter to display the polarity of your reading. You may have the negative lead on the positive terminal and the positive lead on the negative terminal. Doing this with an analog meter might "peg" the needle. The dmm will only indicate the polarity and display a discrete value—provided the range switch is set correctly.

A 100% overrange simply indicates that if the highest range is 1000, the meter will indicate a value as high as 1999.

The autoranging feature enables you to set the dmm to autoranging and the meter will automatically indicate the value within the design limits. Autoranging is usually included in higher-priced models.

Q3-32. A digital multimeter has a _____ switch(es) and a _____ switch(es).

Q3-33. The most-significant digit of a 4½-digit autoranging digital multimeter will be either a(n) _____ or a(n) _____.

Q3-34. You must observe polarity with a digital multimeter unless the meter has _____ _____.

MULTIMETER SAFETY RULES AND PRECAUTIONS

Rule 1: **When not in use, always set the selector switches to the highest dc voltage position.**

There are two reasons for this rule. First, as you will learn later, a multimeter contains batteries; at the highest dc voltage position, the batteries are disconnected from the internal circuits and will not be supplying current. Second, this position of the selector switch provides the best protection for the delicate meter movement in the event the probes should accidentally come in contact with an energized circuit.

Rule 2: **When the meter is in use, forget that it is a multipurpose instrument and think of it only in terms of the function for which you are using it.**

A multimeter with its many switch positions and multiple scales can be confusing and can lead even the best technician into making unnecessary errors. Regard the instrument each time as a particular single-purpose meter.

Rule 3: **When measuring any voltage or current, always use the highest range available first.**

This advice not only provides the best protection to the meter, but it also quickly identifies the best range scale you should use. If the quantity being measured on this or any range causes the needle to move past the end of the scale, immediately remove the probe from the circuit.

Q3-35. A multimeter should be stored with the switches in what position?

Q3-36. Make all measurements first at the _____ _____ setting.

WHAT YOU HAVE LEARNED

1. Meters are used to indicate the quantity or value of voltages and currents.
2. Meters are read by noting the position of a pointer on a marked scale. Digital meters have numbers displayed.
3. Voltmeters are used to measure voltage.
4. Ammeters are used to measure current.
5. Dc meters should not be used to measure ac, and ac meters should not be used for dc.
6. The range of a meter is indicated by the highest marking on the scale. The range is read as "zero to some number." For example, 0-150 volts dc.
7. Never connect a meter to measure a quantity known to be above the meter range. Meter damage will result. You should have some idea of the maximum value of the quantity before making the measurement.
8. A multimeter is a multipurpose meter. A typical instrument will measure ac volts, ac amperes, dc volts, dc amperes, and ohms. It will measure each of these functions in several ranges.
9. Voltage measurements are made by connecting the voltmeter probes across the terminals of the voltage source or device to be measured.
10. If dc voltage is being measured, observe the polarity rule. The terminals of the meter and the circuit should be connected negative to negative and positive to positive.
11. Current measurements are made by connecting the ammeter into the circuit in a manner which allows the circuit current to flow through the meter. This normally requires breaking the circuit and connecting the ends to the meter terminals.
12. If dc current is being measured, observe the polarity rule. Connect the ammeter into the circuit in a manner

which allows current to enter the negative terminal of the meter.

13. Switches are provided on the front panel of a typical multimeter. An ac-dc switch prepares the meter for the type of voltage or current to be measured. A function selector switch sets the meter to the function (volts, amps, ohms) to be measured and the desired range.

14. When not in use, a multimeter should be set at its highest dc-voltage position.

15. When working with a multimeter, forget its many purposes. Think only of the specific function for which you are using it.

16. When measuring voltage or current, always use the highest range first. Remove the probes immediately if the pointer moves past the end of the scale.

17. In many respects using a digital multimeter is similar to using an analog meter. There is a function switch and a range switch. Some digital multimeters have an automatic polarity capability. The immediate display of a discrete number is the principle physical difference. However, the usage rules for analog meters should be followed when using a digital multimeter.

4

The Basic
Telephone System

what you will learn

You will find that a telephone system is a simple electrical circuit which operates in accordance with the principles you have learned earlier. Parts of the telephone circuit convert sound into electrical signals. Other parts change the electrical signals back into sound. As a result, conversations can be transmitted through wires for extremely long distances.

You are familiar with the mouthpieces and earpieces of a telephone. When you finish this chapter, you will understand how these parts work. You will also learn how they are connected in an operating system.

THE MECHANICAL TELEPHONE

Have you ever built a mechanical telephone using a pair of tin cans and a length of string? If you have, you know sound can be transmitted through a string. As crude as this mechanical system is, it demonstrates many of the principles used in the modern telephone.

Q4-1. Vibrations, representing sound, travel down the string of a(an) _____ telephone system.

Q4-2. _____, representing sound, travels down the wires of an electrical telephone system.

The illustrations in Fig. 4-1 show how a mechanical telephone system operates.

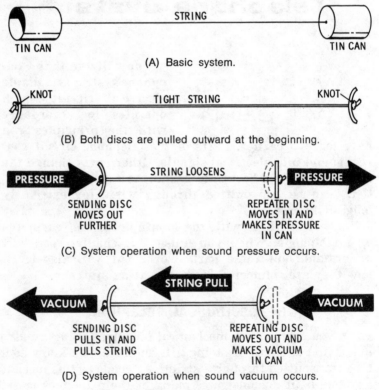

(A) Basic system.

(B) Both discs are pulled outward at the beginning.

(C) System operation when sound pressure occurs.

(D) System operation when sound vacuum occurs.

Fig. 4-1. The mechanical telephone.

Note that the key part of the mechanical telephone system is the flexible metal disc at the bottom of each can. Speaking into the can causes sound waves to strike the disc and make it vibrate.

The vibrations from one disc are carried to the other disc by a tightly stretched string. The second disc repeats the in-and-out motions of the sending disc and develops varying air pressures in the can. These are sound waves which are crude reproductions of the original sound waves.

Principles of Sound

A study of the basic principles of sound reveals how a mechanical (or electrical) telephone system works. Sound is made up of vibrations. Differences in sound are determined by their **frequency**—a number of times per second a sound vibrates. A high tone (a shriek) has a high frequency—several thousand vibrations per second. A low tone (deep bass voice) has a frequency of only a few hundred vibrations per second.

Each tone has a specific frequency (referred to as its pitch). A tuning fork, for example, vibrates and creates a sound tone at the frequency for which it was designed. The same is true of piano or violin strings, the skins of a drum, or your vocal cords.

Air consists of a large number of extremely tiny particles, several million per cubic inch. When sound causes these particles to vibrate, they alternately pack together and fly apart at the frequency of the sound. Packing together creates instantaneous areas of high pressure and flying apart develops a condition of less-than-normal pressure.

As the areas of changing air pressure strike other adjacent air particles, the process is continued. This is the manner in which sound travels through air. When the changes in air pressure strike a flexible disc (or **diaphragm**), it vibrates. The vibrations are at the same frequency as the original sound.

In the mechanical telephone, the sending disc transmits its vibrations to a tightly stretched string which, in turn, sets up the same vibrations in the receiving disc. In the modern telephone, proper design and the use of electricity result in excellent reproduction of sound.

Q4-3. Frequency of a sound indicates the number of times it will _____ in a _____.

Q4-4. Sound vibrations set up corresponding changes of air _____.

THE ELECTRICAL TELEPHONE SYSTEM

The basic telephone system consists of a **mouthpiece** connected to an **earpiece** by electrical wires. This system permits conversation in one direction only. For two-way conversation, each end of the system requires a mouthpiece and an earpiece.

Sound Into Electricity

There are two basic methods of converting sound into electric signals. One method causes current already flowing in a circuit to vary in accordance with the frequency of the sound. The other method converts sound into a varying voltage which, in turn, causes a varying current to move through the circuit. How each method is accomplished will be discussed further.

When the current signals (varying at the rate of the sound) arrive at the earpiece, the process is reversed. If the signal is the type superimposed on an existing current, the variations are received by a material that expands and contracts with the signal frequency. If the mouthpiece develops a voltage to cause a fluctuating current, the current develops a similar voltage in the earpiece.

Basic Parts of a Telephone

The working parts of a telephone mouthpiece (often called a **transmitter**) and an earpiece (sometimes called a **receiver**) are usually identical. The mechanism or material used to produce a varying current depends on the method used. Since all vibrations enter or leave the telephone as changes in air pressure, the transmitter and receiver both contain a diaphragm. Other parts of the mouthpiece include wires, terminals, and materials to hold the parts together.

Connections

In a large telephone system, several lines are connected through a switchboard.

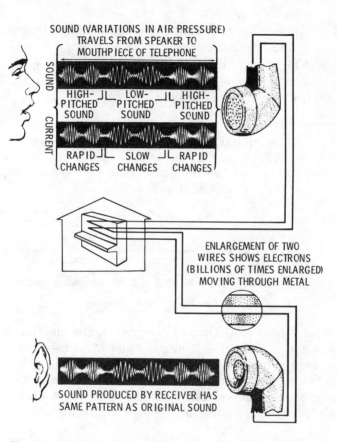

SOUND (VARIATIONS IN AIR PRESSURE) TRAVELS FROM SPEAKER TO MOUTHPIECE OF TELEPHONE

SOUND

CURRENT

HIGH-PITCHED SOUND — LOW-PITCHED SOUND — HIGH-PITCHED SOUND

RAPID CHANGES — SLOW CHANGES — RAPID CHANGES

ENLARGEMENT OF TWO WIRES SHOWS ELECTRONS (BILLIONS OF TIMES ENLARGED) MOVING THROUGH METAL

SOUND PRODUCED BY RECEIVER HAS SAME PATTERN AS ORIGINAL SOUND

Fig. 4-2. Electricity carries the pattern of sound from one telephone to another

Q4-5. A(an) _____ connected to a(an) _____ by wires is the simplest telephone system.

Q4-6. _____ in a telephone line varies at the frequency of the original sound.

Q4-7. Vibrating air pressure strikes a(an) _____ in the (mouthpiece, earpiece).

COMMERCIAL TELEPHONES

As you can see in the illustration in Fig. 4-3, a telephone system is actually a simple circuit.

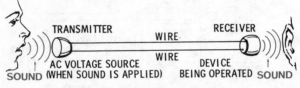

Fig. 4-3. A telephone circuit.

In most commercial (home) telephones, the mouthpiece and earpiece are contained in a single handset. An exploded view of the main parts of a handset is shown in Fig. 4-4.

The transmitter contains a diaphragm resting against **carbon granules** (grains). The granules are loosely packed with enough freedom to expand and contract in volume.

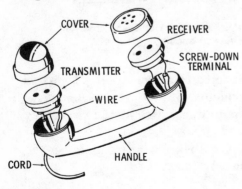

Fig. 4-4. A telephone handset.

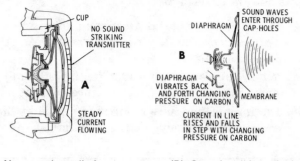

(A) No sound applied.   (B) Sound striking diaphragm.

Fig. 4-5. Transmitter operation.

Transmitter Operation

Part A in Fig. 4-5 shows a drawing of the transmitter with no sound applied. Part B demonstrates what happens when sound strikes the diaphragm.

When the handset is lifted from its cradle, a steady current from the phone system starts to flow through the carbon granules. Sound striking the diaphragm places a varying pressure on the carbon. When the granules are packed tightly, current in the circuit increases. When the diaphragm releases its pressure, the granules become loose and less current flows.

The transmitter diaphragm vibrates in response to the frequency of the sound. Packing and loosening of the carbon follow the vibrations of the diaphragm, and current in the phone line varies with the density of the carbon. Therefore, the current varies at the same rate (frequency) as the original sound.

In a commercial system, the varying current is routed to the desired receiver through a central telephone office.

Q4-8. When a commercial telephone handset is lifted from its cradle, a(an) _____ _____ flows through the transmitter.

Q4-9. When the carbon granules in the transmitter become more densely packed (more, less), current flows.

Q4-10. At what rate does the current vary in the phone lines during a conversation?

Receiver Operation

The method of converting the current back into sound is slightly different. Part A in Fig. 4-6 shows a typical receiver used in a commercial phone system.

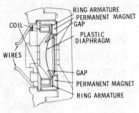
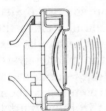

(A) Typical receiver.

(B) Diaphragm vibrates when sound strikes it.

Fig. 4-6. Receiver operation.

Part B shows that the diaphragm vibrates when a sound-varying current passes through the receiver. The current passing through the coil develops a magnetic field which varies in strength with the changes in current. Thus, the field developed by the current periodically repels and attracts the steady magnetic field of the permanent magnet, causing the magnet to which the diaphragm is fastened to move back and forth. This action reproduces the original sound.

SOUND-POWERED TELEPHONES

Another application of a simple electrical circuit is the sound-powered telephone system. It is used only for short distances because of its limited range.

Induced Current

Transmission takes place in sound-powered phones because of the ability of a magnet to **induce** current in a coil of wire. A bar magnet has a north (N) and a south (S) pole. Actually,

Fig. 4-7. Bar magnet has north and south pole.

when the magnet is suspended in air, the north end of the magnet points toward the north geographic pole. Such a magnet has a magnetic field existing in the space surrounding it, with the lines of magnetic force taking the directions shown.

The illustration in Fig. 4-8 shows a bar magnet being moved back and forth inside a wire coil. As the magnet moves, its magnetic lines of force cut across the turns of the coil. This causes an induced current which will actually flow if the ends of the coil are connected to a circuit.

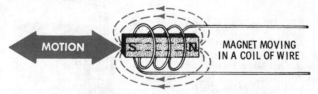

MOTION

MAGNET MOVING IN A COIL OF WIRE

Fig. 4-8. Inducing a current in a coil of wire.

The current reverses direction each time the motion of the magnet changes direction. The amount of current that flows depends on the strength of the magnetic field, the number of turns in the coil, and the speed at which the magnet is moving. Increasing any of these factors increases the amount of current.

Q4-11. Current flowing in phone lines at the time when no sound is present is (dc, ac).

Q4-12. Current flowing during sound transmission is (dc, ac).

Q4-13. What causes a receiver diaphragm to vibrate?

Q4-14. A magnet is surrounded by magnetic _____ of _____.

Q4-15. A magnet moving inside a coil _____ current in the coil.

Your Answers Should Be:

A4-11. Current flowing in phone lines at the time when no sound is present is **dc.**

A4-12. Current flowing during sound transmission is **ac.**

A4-13. A varying current develops a changing magnetic field which repels and attracts a magnet connected to a diaphragm.

A4-14. A magnet is surrounded by magnetic **lines** of **force.**

A4-15. A magnet moving inside a coil **induces** current in the coil.

The induced current theory can be proved by performing the experiment shown in Fig. 4-9. A few microamps of current will be developed.

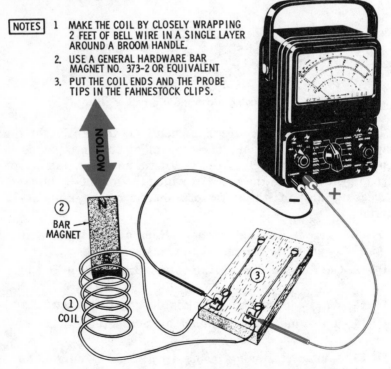

NOTES
1 MAKE THE COIL BY CLOSELY WRAPPING 2 FEET OF BELL WIRE IN A SINGLE LAYER AROUND A BROOM HANDLE.
2. USE A GENERAL HARDWARE BAR MAGNET NO. 373-2 OR EQUIVALENT
3. PUT THE COIL ENDS AND THE PROBE TIPS IN THE FAHNESTOCK CLIPS.

MOTION

② BAR MAGNET

① COIL

③

Fig. 4-9. Creating voltage and current with a magnet.

Sound-Powered Transmitter

The transmitter of a sound-powered phone makes use of the induced-current principle.

As shown in Fig. 4-10, a bar magnet is fixed to the center of a diaphragm. The diaphragm is fastened to the transmitter in such a way that the magnet is over the center of the coil.

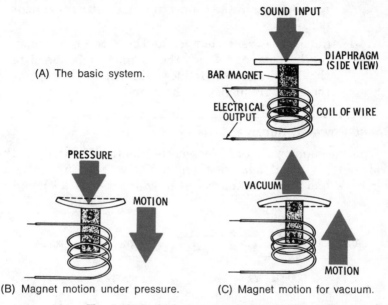

(A) The basic system.

(B) Magnet motion under pressure.

(C) Magnet motion for vacuum.

Fig. 4-10. A diaphragm-operated magnet.

As in other phones, this diaphragm vibrates at the frequency of the sound waves striking it. As it vibrates, the bar magnet moves back and forth within the coil. This induces a current which changes direction at the same frequency as the sound. If the transmitter is connected to a sound-powered receiver, current will flow back and forth through the circuit.

Q4-16. The amount of current induced in a coil can be increased in two ways. Describe them.

Q4-17. A sound-powered phone system (does, does not) have current flowing in the connecting lines during a silent period.

Q4-18. Why will the induced current in this system change at the same frequency as the sound?

Sound-Powered Receiver

The mechanism in the receiver is identical to that in the transmitter—a coil, a magnet, and a diaphragm. Current flowing back and forth in the receiver coil develops a changing

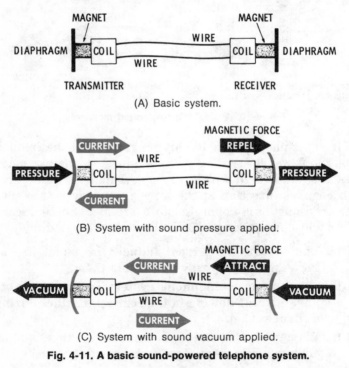

(A) Basic system.

(B) System with sound pressure applied.

(C) System with sound vacuum applied.

Fig. 4-11. A basic sound-powered telephone system.

magnetic field. The bar magnet responds by moving back and forth at the same frequency as the current, causing the diaphragm to reproduce the original sound.

Sound-Powered System

The illustration on the opposite page shows the transmitter and receiver connected together as a working system.

The two bottom figures demonstrate how the induced current changes direction with the motion of the transmitter diaphragm. The corresponding effect upon the receiver diaphragm is also shown.

WHAT YOU HAVE LEARNED

1. A basic telephone system uses all the principles of a simple circuit.
2. The operating principle of any telephone system is the ability of the transmitter and receiver diaphragms to vibrate in unison. The transmitter is the cause and the receiver is the effect.
3. Sound vibrates at a pitch determined by its frequency.
4. Sound vibrations cause changes in air pressure.
5. Changes in air pressure cause a thin metal or plastic diaphragm to vibrate. A vibrating diaphragm also causes changes in air pressure, producing sound.
6. In a commercial phone system, a steady current flows through the circuit during periods of silence. This current is varied when sound waves strike the diaphragm, the resulting vibrations exerting and releasing pressure on carbon granules.
7. At the receiver end, the changing current produces a varying magnetic field which repels and attracts a magnet connected to a diaphragm. This action reproduces the original sound.
8. Another type of phone is the sound-powered system. This type uses induced current for signal transmission.
9. An alternating current is induced in a coil by the back-and-forth motion of a magnetic field.
10. A sound-powered transmitter develops an induced current caused by the vibrations of a diaphragm.
11. A sound-powered receiver uses a changing current to

develop a varying magnetic field about a coil. The field vibrates a magnet connected to a diaphragm, the latter reproducing the original sound.

5

Reading Diagrams

what you
will learn

Before proceeding with more complex circuits, you should learn how to read and draw diagrams used in electricity and electronics. There are many varieties of diagrams, but they have all grown from two basic types—wiring and schematic. The fundamentals of both will be explained.

THE REASON FOR DIAGRAMS

A textbook can be written without illustrations, but very few are. Words alone cannot fully describe the idea or thought the author wants the reader to understand. A writer uses drawings or illustrations with his words to make sure his descriptions are more completely understood.

Most of the illustrations used thus far in this book have been in three-dimensional form. A dry cell was drawn as it actually looks—in the shape of a cylinder and with its terminals in the correct positions. A lamp appeared similar to those in your home. Wires were drawn to look as natural as possible. The artwork was time-consuming but necessary.

However, can you imagine the task required to draw all of the circuits, parts, wires, and terminals of a television set in three-dimensional form?

The illustrations would not only be difficult to draw, but also awkward to use. Technical drawings are needed by engineers who design equipment, workers who construct it, and technicians who service it. They are also required by persons who study electricity and electronics.

Two-Dimensional Diagrams

Two-dimensional (flat instead of shaped) diagrams are now used almost exclusively because they are easier to "read." Reading a diagram means obtaining information from it, such as following the path of current through a circuit. Reading a two-dimensional diagram is simplified by eliminating unnecessary and confusing details. But reading this type of diagram is easy only if you understand the language.

The Language of Symbols

Electrical and electronic diagrams have symbols that either resemble or represent the real item. There are symbols (most of them rather simple) for every electrical or electronic part. When new parts are invented—such as the transistor—a corresponding identifiable symbol is also developed.

Using symbols instead of cumbersome pictures is not new. Shortly after man emerged from the Stone Age, he found it difficult to work with his counting and numbering system. Making marks on the ground or stacking pebbles in a pile became fairly tedious when he wished to indicate "how many" of anything. Numeral symbols were invented to show how many. This permitted the ancient Arab who owned nine sheep and four horses to show on his inventory record the symbols "9" and "4" instead of a number of marks.

Learning electrical and electronic symbols requires the same process you used to learn the meaning of numerals. Learn what the symbols stand for and how to use them.

WIRING DIAGRAMS

Wiring diagrams are used as a guide when constructing a circuit or equipment. They are also useful for locating wires or connections when servicing or troubleshooting.

Basic Wiring Diagrams

The fundamental wiring diagram shows a symbol for each part. Emphasis is placed on displaying the terminals of each part and the wire connections between them. A circuit can be easily put together by following a wiring diagram. As an example, compare the two diagrams in Fig. 5-1.

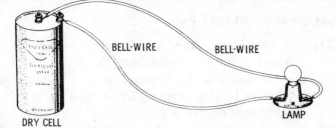

(A) Three-dimensional pictorial diagram.

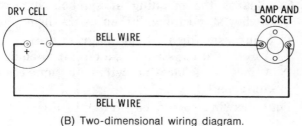

(B) Two-dimensional wiring diagram.

Fig. 5-1. Lamp circuit diagrams.

Note how easy it would be to follow the two-dimensional diagram if you were to construct the circuit. The symbols are easily identifiable. The dry cell is a flat circle (top view) with terminals in the correct positions and polarity markings shown. The lamp symbol is two circles, a bulb in a base, plus two terminals. Wires are straight lines, and the parts are labeled.

Q5-1. The two basic types of electrical/electronic diagrams are _____ and _____.

Q5-2. Why are two-dimensional diagrams used in electricity and electronics? Give two reasons.

Q5-3. Reading a technical diagram requires an understanding of _____.

Q5-4. How does one learn the meaning of symbols?

Q5-5. Name two purposes of a wiring diagram.

Q5-6. Redraw the wiring diagram at the top of the page showing a knife switch placed in the circuit between the lamp and cell. Draw a side view of the switch.

Your Answers Should Be:

A5-1. The two basic types of electrical/electronic diagrams are **wiring** and **schematic.**

A5-2. Two-dimensional diagrams are used because they are **easier to draw** and **easier to read** than a three-dimensional diagram.

A5-3. Reading a technical diagram requires an understanding of **symbols.**

A5-4. One learns the meaning of symbols by learning what they stand for and then using them.

A5-5. Two purposes for a wiring diagram are:
1. **A guide for constructing a circuit or equipment.**
2. **A means of locating wires or connections in equipment.**

A5-6. Your drawing should be similar to this:

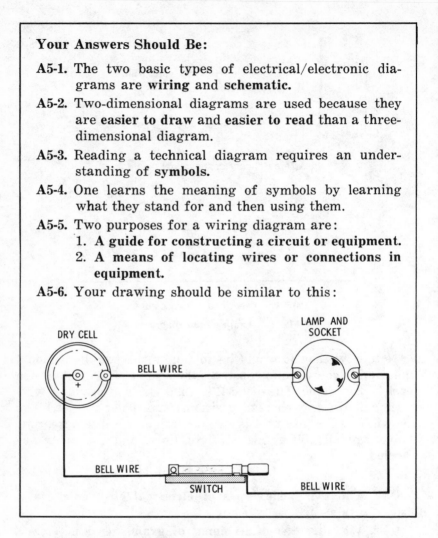

No two manufacturers will necessarily use identical symbols for the same part. Each symbol, however, will be a close representation of the real thing. The switch symbol, thus, should show the terminals and make clear the difference between the open and hinged ends.

The illustration in Fig. 5-2 shows a wiring diagram for two 1.5-volt lamps connected to a 1.5-volt dry cell. You may construct it if you wish. The circuit is intended to demonstrate a principle of electricity.

You will note that the lamp symbol has been changed. A single circle is often used for this purpose.

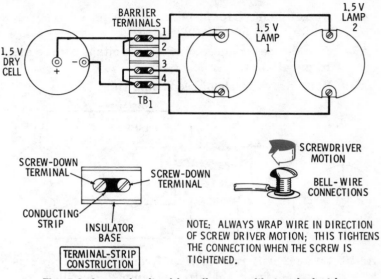

Fig. 5-2. Lamp-circuit wiring diagram with terminal strip

A new part has been added—a terminal board (TB1). As shown in its construction detail, a terminal board has a metal strip with two screws on either end, mounted on an insulating material. These boards serve as connecting points for wires.

The electrical principle demonstrated in the diagram above is that two lamps can be connected to a single voltage source. Since both lamps are connected **across** the voltage source (the top terminal of each lamp is wired to the positive pole and the bottom terminal to the negative pole), the lamps are said to be in **parallel.** This means the same voltage (1.5 volts) is being applied to each lamp.

Q5-7. What is the purpose of a terminal strip?

Q5-8. Two devices connected across a voltage source are in _____.

Q5-9. Redraw the circuit in Fig. 5-2, showing by means of arrows the direction of current flow in each wire.

Q5-10. Remove lamp 1 from the circuit by disconnecting the wire at terminal 3. Will lamp 2 go out?

A5-7. A terminal strip provides a means of securely joining two or more wires.

A5-8. Two devices connected across a voltage source are in **parallel**.

A5-9. Your drawing should look like this:

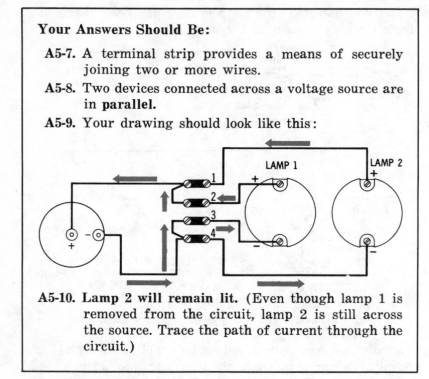

A5-10. **Lamp 2 will remain lit.** (Even though lamp 1 is removed from the circuit, lamp 2 is still across the source. Trace the path of current through the circuit.)

Multiwire Diagrams—The wiring diagram in Fig. 5-3 shows three lamps in parallel.

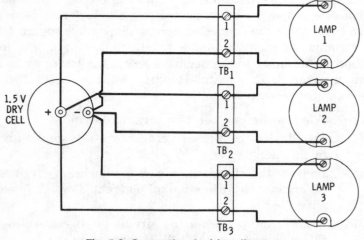

Fig. 5-3. Conventional wiring diagram.

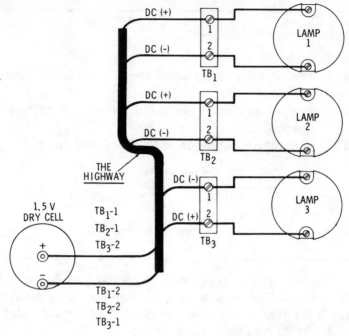

Fig. 5-4. A highway wiring diagram.

Another Multiwire Diagram—With only three lamps the diagram is cluttered. However, another type of wiring diagram removes the clutter. Since it looks like a highway with secondary roads leading from it you might call it a highway wiring diagram. The same three lamps are redrawn in Fig. 5-4. Each wire entering the "highway" has its destination marked.

The abreviation for terminal board 1 is TB1. Since TB2-1 (terminal 1 of terminal board 2) is positive it is connected to the positive pole of the cell. Although only two wires are shown connected to the cell, there are actually three at each pole as indicated by the TB listings.

Q5-11. To which pole of the dry cell is the wire from TB3-2 connected?

Q5-12. Which dry-cell terminal is connected to TB2-2?

Q5-13. What would the abbreviation TB4-3 mean?

Q5-14. Draw a wiring diagram showing only lamp 4 connected to TB4 with TB4-1 positive.

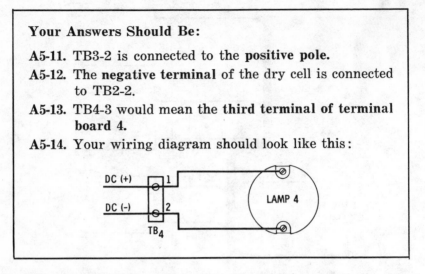
The Airline Wiring Diagram—The airline wiring diagram shows wire destinations without a connection between terminals. This type of diagram is used in the same manner as the highway diagram.

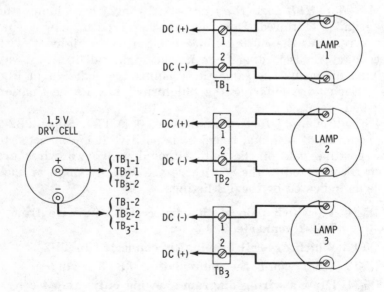

Fig. 5-5. An airline wiring diagram.

SCHEMATIC DIAGRAMS

Schematic diagrams are used more than any other technical diagram in electronics. Engineers use **schematics** (the term "diagram" is usually dropped) when designing equipment and testing its performance after construction. Technicians constantly refer to a schematic while servicing or troubleshooting equipment.

Information including a schematic is available for nearly every television set, radio, and other electronic equipment ever manufactured. These can be purchased at electronic supply stores and from mail order companies—the same source from which you purchase electronic components to repair or construct equipment.

The schematic is used in nearly all electricity/electronics textbooks. The reason for using this kind of diagram, of course, is the need for all future technicians and engineers to become familiar with the type of diagrams they will be using most often. Another reason is the clarity with which the schematic provides information. The many parts of a circuit, or group of circuits, can be drawn in a limited amount of space. The symbols used are fairly standard and do not vary as the representations do in wiring diagrams.

Symbols

Shown in Fig. 5-6 is the schematic symbol for a lamp.

(A) Schematic.

(B) Pictorial.

Fig. 5-6. Symbols for a lamp.

Q5-15. What is the difference between highway and airline diagrams?

Q5-16. Technicians use schematic diagrams to _____ and _____ equipment.

Q5-17. Engineers use schematics to _____ and _____ equipment.

Q5-18. What does the curved line inside the symbol for a lamp represent?

Your Answers Should Be:

A5-15. A highway wiring diagram has a broad line (highway) drawn to each of the wires in the circuit. An airline diagram does not.

A5-16. Technicians use schematic diagrams to **service** and **troubleshoot** equipment.

A5-17. Engineers use schematics to **design** and **test** equipment.

A5-18. The curved line inside the symbol for a lamp represents its **filament**.

Cells and Batteries—The symbol for a cell is two broad parallel lines, one shorter than the other, each with a perpendicular line attached. The broad lines represent the negative and positive materials (plates) in a cell. Since a battery contains two or more cells, its symbol is two or more pairs of plates—the standard symbol usually used is either two or three pairs. Since a battery may have any number of volts, the symbol is usually labeled with the voltage. It is also good practice to mark the polarity of the battery on the symbol, (−) for negative and (+) for positive.

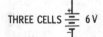

ONE CELL 1.5V

FLASHLIGHT CELL (1.5 V)

TWO CELLS 3V

3-VOLT BATTERY

THREE CELLS 6 V

AUTO BATTERY

Fig. 5-7. Symbols for cells and batteries.

A Circuit Schematic—Now that you are familiar with the symbols for a voltage source and an operating device, you should be able to draw the schematic of a simple circuit. It should look like Fig. 5-8.

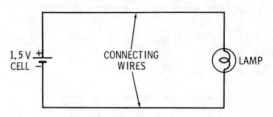

Fig. 5-8. A circuit schematic.

The lamp and cell symbols are connected by lines representing wires. Note that the lines run in only two directions— horizontal and vertical. Slanted or curved lines lessen the clarity of a diagram. Also note that a voltage value and polarity markings appear on the cell.

Switches—The symbol for a simple switch is also suggestive of the original article.

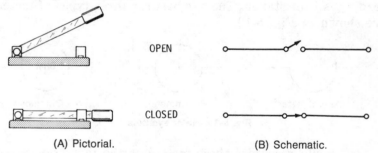

(A) Pictorial.   (B) Schematic.

Fig. 5-9. Symbols for a switch.

The arrowhead has no real significance other than helping to identify the symbol as a switch when it is shown in a closed position. Without the arrowhead, the symbol would look like a wire between two terminals.

Q5-19. Draw a schematic of a 6-volt battery and a lamp. Place an open switch in the line attached to the positive terminal and a closed switch in the line to the negative terminal.

Q5-20. Will current flow in this circuit?

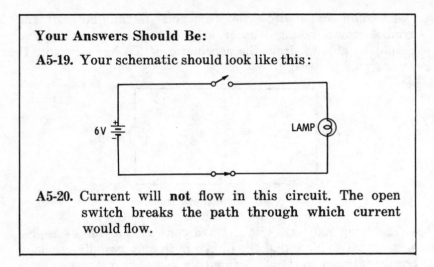
Meters—Meters are quite often inserted into circuits to monitor voltage and current. The **wattmeter** (a combination of a voltmeter and an ammeter) in your home is an example. It is often necessary to show on a schematic where a meter reading is being taken. The symbols for three types of meters are shown in Fig. 5-10.

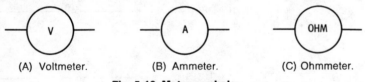

(A) Voltmeter. (B) Ammeter. (C) Ohmmeter.

Fig. 5-10. Meter symbols.

You should recognize these meters as the three types built into a multimeter. They are also available as separate meters.

Voltage Sources—The battery symbol obviously represents a source of dc voltage. There are also other types of dc sources. The symbol for one of these, in addition to symbols for ac sources, is shown in Fig. 5-11.

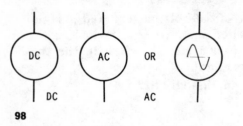

Fig. 5-11. Symbols for ac and dc voltage sources.

Both dc and ac voltages can be supplied by generators or developed by other methods. An example is the ac voltage developed by the vibrating diaphragm of a telephone. The basic symbol is the circle, with the letters dc or ac (to designate the type of voltage) placed inside. The second ac symbol includes a **sine wave** instead of letters. The sine wave in the circle represents the rise and fall of alternating voltage.

Coils—A coil symbol actually looks like the several turns of wire it represents. The symbol on the left in Fig. 5-12 is the most common version.

Fig. 5-12. Schematic symbols for a coil.

Wire Connections and Crossings—Quite often one wire is connected to another. In a wiring diagram the terminal where the connection is made is shown. In a schematic, however, terminals are usually not indicated. In addition, it is sometimes necessary to show lines crossing each other. The illustration in Fig. 5-13 shows how connections and crossings are indicated.

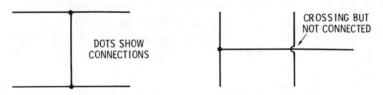

Fig. 5-13. Symbols for connecting and crossing wires.

Series and Parallel Connections—In the discussion on wiring diagrams, two lamps were connected in **parallel**. Both were connected **across** the battery terminals. Devices can also be connected in **series**, like knots in a string. If two lamps are connected in series with a battery, the same current flows through both lamps.

Q5-21. Draw a schematic of two lamps and a 117-volt ac source, all connected in series.

Q5-22. Draw two lamps in parallel across a 12 V dc generator. The second lamp is to be switched out of the circuit.

Your Answers Should Be:

A5-21. Your series circuit should look like this:

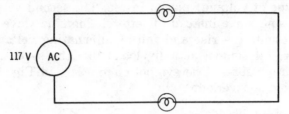

You may have used the other symbol for an ac source.

A5-22. Your parallel circuit should look like this:

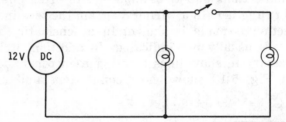

The two lamps are in parallel across the generator. The open switch disconnects the second lamp. With the switch closed, both lamps will light. The dots show the points at which the wires are connected.

There are dozens of other symbols which will be shown at the time new parts or devices are introduced. The purpose of this chapter was to teach you the basic principles of how to read and draw simple diagrams. More complicated diagrams which include many different components will be used later as you gain more experience and become familiar with simple schematics.

As a summary of the fundamentals of schematics, how would you draw this circuit?

1. The voltage source is two 6-volt batteries (series-connected) in parallel with a 12-volt battery. The polarity of all three batteries is in the same direction.

2. The **load** (another term for operating devices) is two 6-volt lamps (connected in series) in parallel with a 12-volt lamp. The load is to be connected across the voltage source.

3. Switches are placed in the circuit so that each parallel **leg** (a separate current path) of the load can be switched on and off individually.

4. An ammeter and a voltmeter are placed in the circuit between the load and source.

The easiest and most accurate way to draw the schematic from the above description is in sections.

1. Section 1—the voltage source.
2. Section 2—the load with its switches.
3. Section 3—the meters between the load and source.

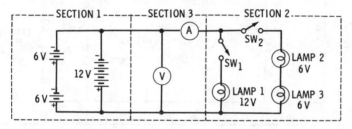

Fig. 5-14. Drawing a schematic by sections.

As you recall, an ammeter is always connected in the circuit path. Therefore, the ammeter in the illustration in Fig. 5-14 is placed in series with the source and the load. A voltmeter is always placed across (in parallel with) the load or source.

Q5-23. Draw a schematic for the following:
1. Load—A coil in series with two coils in parallel. "L" is the symbol for a coil. Label the coils L_1, L_2, and L_3 in the order of their distance from the source.
2. Source—Two 100-volt ac generators in series.

Q5-24. Draw a schematic for two 50-volt dc generators (in parallel) supplying voltage for three 50-volt lamps, also in parallel. Show a voltmeter to measure the source voltage and an ammeter to measure the current through lamp 2.

Your Answers Should Be:

A5-23. Your schematic should be similar to this:

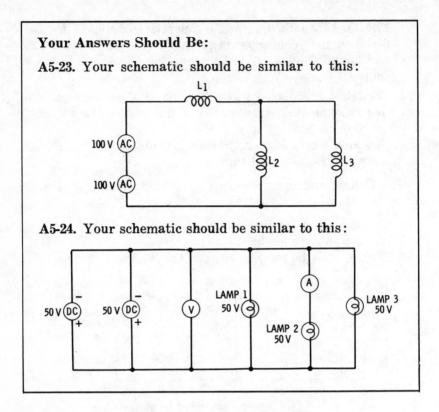

A5-24. Your schematic should be similar to this:

WHAT YOU HAVE LEARNED

1. Technical diagrams are drawn with symbols to clearly present a great deal of information in a limited amount of space.
2. There are two basic types of diagrams generally used in electrical and electronic work. These two types are wiring and schematic diagrams.
3. Wiring diagrams are useful as a guide when detail is required for construction purposes.
4. Schematic diagrams with their simple symbols are widely used by engineers for designing and testing equipment, and by technicians for servicing and troubleshooting.
5. A fundamental or conventional wiring diagram shows each wire and a representative symbol for each part, and clearly labels each terminal.

6. When several parts must be included, an airline or highway wiring diagram is drawn. A highway diagram uses a broad line representing the many wires going to each terminal. All terminals are marked. An airline diagram contains the same details without the broad line.
7. Wiring diagrams usually contain too much detail for general use, other than construction. The most widely used diagram is called a schematic.
8. Symbols for circuit parts have been fairly well standardized. They are simple in detail and represent the actual article.
9. New electrical terms introduced were the circuit relationships of series and parallel connections. Parts are in parallel when their respective terminals are connected to the same point. Parts are in series when they are connected together in line.
10. Symbols were shown for several different parts or devices—lamps, cells, batteries, switches, meters, ac and dc sources, coils, and connecting wires.
11. You were shown how to draw schematics of both simple and complex circuits.

6

Understanding
Resistors

what you
will learn

Voltage, current, and
resistance are closely related within a circuit.
Where you find current,
you find the other two.
Current cannot flow unless there is voltage. How much
will flow is determined by how much voltage and how
much resistance are present in the circuit. You will learn
what resistance is and how to measure it.

WHAT LIMITS CURRENT FLOW?

You have learned that voltage is a pressure which forces
current to flow through a circuit. You also have learned that
current has the ability to heat a lamp filament white-hot and,
thus, produce light. But have you ever wondered why a 40-watt
lamp produces less light than one rated at 100 watts? The
amount of voltage pushing current through both lamps is the
same. The answer, of course, is the individual characteristic
of each lamp which limits the amount of current that will flow.

The 100-watt lamp glows more brightly because more current is allowed to pass through the filament, heating it to a
higher degree, thus causing it to give off more light. Less
current is allowed to flow through the 40-watt filament. The
reason for the different amount of current through each of
the two lamps is an electrical characteristic called **resistance**.
Resistance **limits** or **controls** the flow of current.

WHAT IS RESISTANCE?

Resistance is a physical property of all materials and is directly responsible for the amount of current which will flow through a material with a given voltage applied.

Atomic Structure

All matter is made up of invisible particles called **atoms.** There are over 90 different atoms, or **elements,** as the physicist calls them. One of the features that makes one atom different from another is the number of **electrons** each contains. A hydrogen atom has one electron, an oxygen atom has eight, and an atom of uranium has 92.

You know that current is a flow of electrons and that electrons are made to move by a voltage. This does not mean that an electron leaves the negative pole of a battery and speeds around the circuit to the positive terminal. Instead, there is a general movement or drift of electrons throughout the complete circuit.

A greatly magnified and exaggerated drawing of a length of wire with four atoms is shown in Fig. 6-1—A, B, C, and D. In the shortest possible length of a very thin wire there are actually many millions of atoms.

Fig. 6-1. Electron movement.

Electron Flow

As shown above, electrons orbit about the center of an atom. At the instant voltage is applied, two things happen simultaneously—negative voltage at one end of the wire pushes against the electrons, and positive voltage at the other end of the wire pulls them toward that end. In moving, electrons strike other electrons. One electron is bumped out of atom A, and it in turn pushes another out of atom B. At the positive end, an electron is pulled from atom D and another leaves atom C to replace it. The atoms of some materials give up their electrons more easily than the atoms of other materials.

Resistance of Materials

There is no perfect conductor. Even the best conductors, such as those having silver or copper atoms, resist the pressure to release electrons. On the other hand, the best insulators have atoms which, under conditions of sufficiently high voltage, give up some electrons. The resistance of a material, then, is **determined by its atomic structure.**

The size of the columns in the illustration in Fig. 6-2 shows the comparative resistance of certain materials. Keep in mind that no material is a perfect conductor or a perfect insulator.

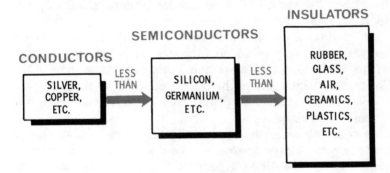

Fig. 6-2. Resistance of materials.

Most metals contain atoms that release electrons very easily. These materials, therefore, offer the least resistance to current flow. Insulators have the greatest resistance because their atoms resist the release of electrons. The in-between materials are neither good conductors nor insulators. Among these are certain materials, the **semiconductors,** from which **transistors** are manufactured.

Unit of Measurement—Resistance is measured in **ohms.** The resistance of the 1.5-volt lamp used in a preceding experiment, for example, is approximately 6 ohms. In other words, the lamp offers 6 ohms of resistance to the electrical pressure of the 1.5-volt cell, and the result is a current flow of 0.25 amp.

Q6-1. What is the difference between a conductor and an insulator?

Q6-2. The resistance of a material is determined by its _____ structure.

Volts, Ohms, and Amperes

Since resistance limits the amount of current that flows and voltage forces an amount to flow, there must be some numerical relationship between them.

You would see current decrease to half its former amount when a second lamp is added in series with the lamp circuit. Current is divided by two when resistance is multiplied by two. Mathematicians say, then, that current is inversely proportional to resistance. In other words, **current decreases as resistance increases.**

You can also discover, by experimenting, what happens to current when voltage increases. You will find they increase together (they are directly proportional to each other). This makes sense because the pressure of voltage causes current to flow. If the pressure increases, flow increases.

These relationships of voltage and resistance to current can be expressed in an arithmetic statement as

$$\text{Current in a circuit} = \frac{\text{voltage applied to a circuit}}{\text{resistance of a circuit}}$$

Using mathematical symbols, this statement becomes

$$I = \frac{E}{R}$$

where,

I is the current in amperes,
E is the voltage in volts,
R is the resistance in ohms.

If voltage (**E**) is increased, current (**I**) will increase. When **E** decreases, **I** also decreases. The relationship between **I** and **R** is just the reverse. A decrease in **R** causes **I** to increase. The larger **R** becomes, the smaller **I** will be.

RESISTORS

Now that you know what resistance is, you are ready to learn about a device called a **resistor**.

What Is a Resistor?

A resistor not only offers resistance to current flow but also has a specific value of resistance. A resistor has many uses, but its main purpose is to control current.

Making a Simple Resistor—Since you know that all materials have some resistance, you should be able to make a resistor having a desired value. The wire from a heating element is an economical material to use.

The element with its wire can be purchased from hardware or most variety stores. Unstretched lengths of the coil wire can be obtained at an appliance repair shop. Replacement wire for heating elements is tightly coiled and looks very much like a spring. If you make the resistance board shown in Fig. 6-3 using the tightly coiled wire, cut off a 2- or 3-inch length. Grasp the last turn on each end with pliers and gently stretch the coil until it is a foot long.

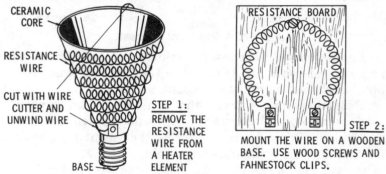

CERAMIC CORE

RESISTANCE WIRE

CUT WITH WIRE CUTTER AND UNWIND WIRE

BASE

STEP 1: REMOVE THE RESISTANCE WIRE FROM A HEATER ELEMENT

RESISTANCE BOARD

STEP 2: MOUNT THE WIRE ON A WOODEN BASE. USE WOOD SCREWS AND FAHNESTOCK CLIPS.

Fig. 6-3. Making a simple resistor.

Q6-3. The unit used in measuring resistance is the _____.

Q6-4. When resistance remains the same, current will (increase, decrease) when voltage decreases.

Q6-5. With E constant, I will (increase, decrease) if R decreases.

Q6-6. Fifty volts is applied across 100 ohms. How much current will flow through the resistance?

Your Answers Should Be:

A6-3. The unit used in measuring resistance is the **ohm**.

A6-4. When resistance remains the same, current will **decrease** when voltage decreases.

A6-5. With **E** constant, **I** will **increase** if **R** decreases.

A6-6. Fifty volts is applied across 100 ohms. The current that will flow is

$$I = \frac{E}{R} = \frac{50 \text{ volts}}{100 \text{ ohms}} = 0.5 \text{ amp}$$

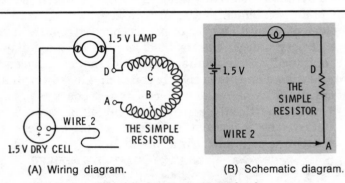

(A) Wiring diagram.　　　　(B) Schematic diagram.

Fig. 6-4. A lamp control circuit.

A Lamp-Control Circuit—The resistance board can be used as a means of controlling current in the familiar lamp circuit. Note the schematic symbol for resistance.

A Potentiometer—A device that can be adjusted to provide a desired amount of resistance is called a **potentiometer**. It has a moving contact that performs the function of moving wire 2 across the resistance from point A to D.

The diagram shows how the simple resistor would look if it were converted into a potentiometer.

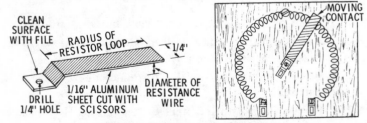

Fig. 6-5. Making a potentiometer.

By inserting the potentiometer into a circuit with one wire connected to the rotating arm and the other to one of the terminals, a desired value of resistance can be selected.

As you can see, the schematic symbol for a potentiometer (variable resistor) is a combination of switch and resistor symbols. The arrow indicates that the value of resistance can be varied.

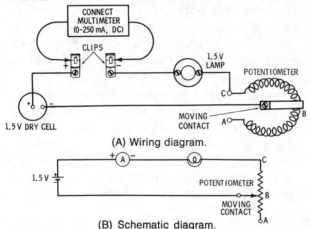

(A) Wiring diagram.

(B) Schematic diagram.

Fig. 6-6. A lamp-control circuit with potentiometer and ammeter.

Q6-7. A potentiometer is a _____ resistor.

Q6-8. Draw the symbol for a potentiometer.

For the remaining questions, use these data:

(a) The lamp and potentiometer are the circuit load.

(b) Lamp resistance is 6 ohms; potentiometer, 12 ohms.

(c) The resistance from point B to point C (see illustration above) is half the potentiometer resistance.

Q6-9. What is the total resistance of the load?

Q6-10. Circuit current will be minimum when the moving contact is at point (A,B,C).

Q6-11. The lamp will glow brightest when the contact is at point (A,B,C).

Q6-12. How much current will be registered by the ammeter when the contact is at point B?

MEASURING RESISTANCE

Like voltage and current, resistance can be measured with a meter. In fact, you have already learned that an **ohmmeter** is part of a multimeter.

The Ohmmeter Scale

An ohmmeter scale is labeled either *ohms* or with the Greek letter omega (Ω). Instead of writing the word "ohm" after the numerical value of a resistance, the omega symbol is often used. A typical *ohms* scale is shown in Fig. 6-7.

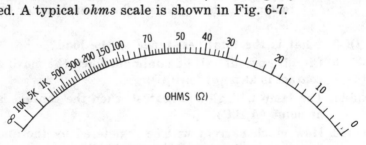

Fig. 6-7. An ohmmeter scale.

Meter Scale—The meter scale reads from zero to infinity, as indicated by the symbol ∞. Because it is so great, infinity has no numerical value. The "K" on the scale stands for 1000; therefore, 5k equals 5000 ohms.

Zero is on the right of the scale instead of the left as it is for voltmeter and ammeter readings. As you remember, the meter pointer moves across the dial a distance that is proportional to the amount of current flowing through the meter. If there is zero current when taking a current or voltage measurement, the pointer remains at the left on 0. Maximum voltage or current readings cause the pointer to rest on the right end of the respective scales.

How the Meter Measures—The ohmmeter measures the value of a resistance by passing current through the resistance. The amount is measured by the meter. The smallest resistance allows the most current to flow. Therefore, zero ohms is at the maximum position of pointer swing—at the right. Maximum resistance permits the least meter current to flow. Therefore, ∞ is on the left.

NOTE: As mentioned previously, there are no resistances that are perfect insulators or perfect conductors. Zero and ∞ have been selected arbitrarily as the two extreme scale markings. Each scale must have a maximum and a minimum point. These points are 0 to ∞.

Reading the Scale—The ohmmeter scale is read in the same manner as the voltage and current scales. If the pointer stops on a numbered division, that number represents the value in ohms. Between numbers, you determine the value of each division mark. Multiply this by the number of marks to the pointer and add to the lower number. For example, there are five divisions between 40 and 50. The pointer rests on the third division past 40. The entire distance is 10, so each division is worth 2. Three 2's are 6 which, when added to 40, provide a meter reading of 46 ohms. Near the left (upper) end of the scale, numbers are close together. A reading here can only be an estimate.

Q6-13. **The pointer rests on the second of five divisions between 500 and 1k. What is the value of resistance?**

Calibrating the Ohmmeter

The ohmmeter must be calibrated, or "zeroed," before accurate resistance measurements can be made.

The zeroing procedure is as follows:

1. Set the function selector to R × 1.
2. Touch the meter probes together. The pointer will rest close to, but not at 0 on the scale.
3. Slowly vary the OHMS ADJ control until the pointer rests on 0.

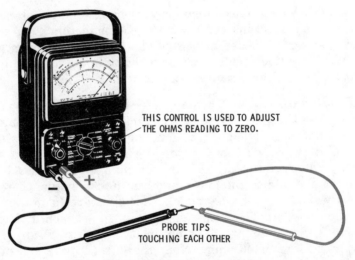

THIS CONTROL IS USED TO ADJUST THE OHMS READING TO ZERO.

PROBE TIPS TOUCHING EACH OTHER

Fig. 6-8. Calibrating the ohmmeter.

The ohmmeter is now zeroed and a resistance measurement can be made. The pointer should swing back to infinity when the meter is not in use and the probes are not touching each other.

Resistance Measurements

Resistance measurements are made by touching the meter probes to the terminals of the unit to be measured. **Never make a resistance measurement when the circuit under test is operating.** Current from the operating circuit will enter the test leads and damage the meter.

The schematic diagram in Fig. 6-9 shows how an ohmmeter is used to measure the resistance of a potentiometer.

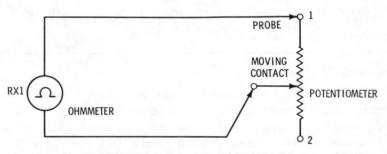

Fig. 6-9. Measuring the potentiometer resistance.

If the potentiometer contact is at terminal 1 and the meter has been properly zeroed, the resistance reading should be zero ohms. Leaving the probes connected, the meter will show the gradual increase of resistance as the contact is moved toward terminal 2. At terminal 2 the total resistance of the potentiometer will be read on the meter scale.

If this experiment is conducted with a properly designed potentiometer, ¼ of the total resistance will be read when the contact travels ¼ of its total rotation, ½ at the halfway point, and so on. This shows that resistance is proportional to the length of the material. If a 12-inch length of wire measures 12 ohms, a 1-inch piece of the same wire will measure 1 ohm.

Q6-14. **After you set the function selector switch to the proper position, how do you complete the procedure of zeroing an ohmmeter?**

Q6-15. **Before making a resistance measurement in a circuit, what should you determine first about the circuit?**

Q6-16. **One yard of wire measures 2 ohms. How many ohms will 10 yards of the wire measure?**

Hot and Cold Resistance

In some cases a resistance taken with an ohmmeter will not be in the true resistance of the device when it is operating in a circuit. A lamp is a good example. It gives off light because current has raised the temperature of the filament to a white-hot level. The physical structure of most materials is such that their resistances will increase with a rise in temperature.

Copper will increase in resistance proportionally to the change in temperature; copper, then, is said to have a positive temperature coefficient. In materials where heat is a factor, there is a hot and cold resistance.

Cold Resistance—Cold resistance is an ohmmeter measurement. It is taken when operating current is not passing through the device and, therefore, not heating the resistance. Cold resistance cannot be used to determine current that will be drawn by a heat-generating device.

Hot Resistance—Hot resistance is the true operating resistance of the heated wire or material. It is this resistance that determines the amount of current flow. Because current is determined by the hot resistance, and because current and resistance are inversely proportional to each other, the value of hot resistance can be determined by

$$\text{Hot (operating) resistance} = \frac{\text{voltage applied}}{\text{current flowing}}$$

You should note that this is just another way of stating the formula, current equals voltage divided by resistance.

Determining Hot Resistance

Since you must know voltage and current values to determine hot resistance, use the method in Fig. 6-10.

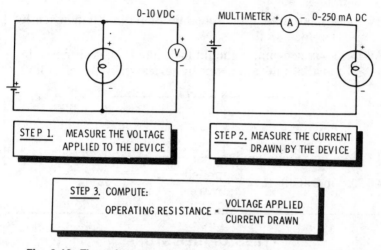

Fig. 6-10. The voltmeter-ammeter method of measuring resistance.

This method of measuring resistance is called the **voltmeter-ammeter method.** Using this method you will find that the dry cell measures very close to 1.5 volts and the current very near 200 milliamps (0.2 amp). In this case, the resistance will be 7.5 ohms. The arithmetic statement says that if you divide volts by amperes the answer will be the value of the heated resistance in ohms.

Using the cold-resistance method (ohmmeter measurement) you would probably find the lamp resistance to be 6 ohms or less.

Q6-17. Cold resistance is usually (higher, lower) than hot resistance.

Q6-18. A heated wire will pass (more, less) current than the same wire when it is cold.

Q6-19. Redraw the two schematics above into one to show the voltmeter-ammeter method with both meters in the circuit at the same time.

Q6-20. If the ammeter reads 0.5 amp and the voltmeter 9 volts, what is the hot-resistance value?

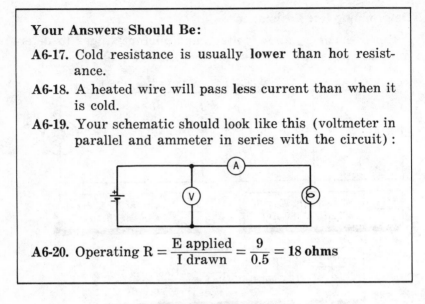

Your Answers Should Be:

A6-17. Cold resistance is usually **lower** than hot resistance.

A6-18. A heated wire will pass **less** current than when it is cold.

A6-19. Your schematic should look like this (voltmeter in parallel and ammeter in series with the circuit):

A6-20. Operating $R = \dfrac{E \text{ applied}}{I \text{ drawn}} = \dfrac{9}{0.5} = 18$ **ohms**

TYPES OF RESISTORS

Resistors are classified in two ways: (1) in terms of their construction (wirewound, composition, and film), and (2) in terms of their type or function (fixed, adjustable, variable).

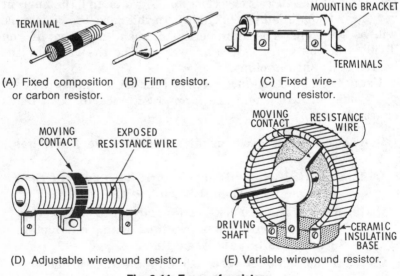

(A) Fixed composition or carbon resistor. (B) Film resistor.

(C) Fixed wire-wound resistor.

(D) Adjustable wirewound resistor. (E) Variable wirewound resistor.

Fig. 6-11. Types of resistors.

Wirewound Resistors

Wirewound resistors are made by wrapping resistance wire around a ceramic or other high-insulation cylinder. The assembly is then covered with enamel glaze and baked. The wire has a known value of ohms per inch.

Composition or Carbon, and Fixed Film Resistors

Composition or carbon resistors are molded from a paste consisting of carbon (a conducting material) and a filler. Terminal wires are inserted into the paste before it hardens. The resistor is then covered with a plastic coating. The resistance and wattage rating of a composition resistor is determined by the ingredients (carbon and filler) and its diameter and length. The resistance of carbon will decrease as its temperature increases; carbon, then, has a negative temperature coefficient.

Film-type resistors use a layer of resistive material on an insulating core. The film thickness determines the resistive value.

Fixed Resistors

A fixed resistor has only one nonvariable ohmic value.

Adjustable Resistors

Adjustable resistors provide a range of resistance within the limits of their total value. When placed in a circuit, the sliding contact can be positioned and secured to accurately provide the required resistance value.

Variable Resistors

Variable resistors are designed for continuous adjustment. A shaft to control the resistance value is usually connected to a knob on a front panel. The volume control of your radio or tv set is an example.

A rheostat is a variable resistor. The material that the moving contact presses against may be either resistance wire or a carbon mixture.

Q6-21. Composition resistors are made from a _____ and _____ mixture.

Q6-22. The control that dims the dashboard lights in an automobile is a(an) _____ resistor.

Your Answers Should Be:

A6-21. Composition resistors are made from a **carbon** and **filler** mixture.

A6-22. The control that dims the dashboard lights in an automobile is a **variable** resistor.

Resistor Applications

Typical applications for each kind of resistor are presented in Table 6-1.

Table 6-1. Resistor Applications

Type	Applications
Composition or Carbon	Composition resistors are the least expensive of the types discussed. They are, therefore, the type most widely used. However, composition resistors have certain limitations. They cannot handle large currents, and their measured values may vary as much as 20% from their rated resistance.
Wirewound	Wirewound resistors are more expensive to manufacture. They are used in circuits which carry large currents or in circuits where accurate resistance values are required. Wirewound resistors can be made to within 99% or better of the desired value.
Fixed Film	Smaller and less expensive than wirewound resistors, film-type resistors have the same accurate resistance values, but not the same large current capability.

RESISTOR POWER RATINGS

As you already know, current passing through a resistor generates heat. If too much heat is generated, the resistor will be damaged. Wire in the wound resistor will melt and become open, or some of the carbon in the composition resistor will burn away.

The current-carrying capacity of a resistor is rated according to the amount of heat it can safely release in a given period of time. A resistor cannot be used in a circuit where current causes heat to build up faster than the resistor can dissipate it. When such a condition exists, the resistor may become so hot that it will be destroyed. Even if the resistor doesn't melt and become open, the excessive heat may cause a permanent change in its resistance value. In addition, heat from the overloaded resistor may damage other components that are near by.

Since heat is a form of energy, the heat-releasing rate of a resistor is measured in energy units. The unit is a watt. A 100-watt lamp dissipates 100 watts of heat. In the process, the lamp also gives off light.

Heat energy depends on the amount of current flowing through a resistor. The arithmetic involved is

Heat energy in watts = (current in amps)^2 × (resistance)

This means that the number of watts dissipated by a resistor can be found by multiplying the resistance in ohms times the square (a number multiplied by itself) of the current in amperes, and is referred to as **power.**

For example, a 10-ohm resistor has three amps flowing through it. What must be its power rating in watts?

Power = (amps)^2 × (ohms) = $(3)^2$ × (10) = 90 watts

Composition resistors usually come in power ratings of $\frac{1}{4}$ watt, $\frac{1}{2}$ watt, 1 watt, and 2 watts. If larger power ratings are required, wirewound resistors are used.

A design engineer determines the value of resistance needed and the amount of current that will flow through it. He then specifies the resistor wattage that must be used. If the value falls between two of the ratings mentioned above, he selects the higher rating.

Q6-23. Which of the standard composition-resistor ratings would you select for a resistor of 10 ohms through which 1/10 of an amp flows?

Q6-24. A 1-watt wirewound resistor (will, will not) safely carry more current than a 2-watt composition resistor.

RESISTOR TOLERANCE

As mentioned previously, a resistor will rarely measure the exact number of ohms specified by its label. The amount it will vary is called **tolerance.** Every resistor has a tolerance rating.

Resistor tolerance is given as a **percentage value** which indicates the amount that a resistor may vary above or below its labeled value. Standard tolerances for composition resistors are 5%, 10%, and 20%. Wirewound resistors may have tolerances as low as 1% or 2%.

Try a 1000-ohm 10% tolerance resistor as an example. Ten percent of 1000 is 100 ohms. The tolerance factor, thus, indicates this resistor will measure somewhere between 100 ohms above and 100 ohms below the labeled value of 1000 ohms. This is a range from 900 to 1100 ohms. The same resistor with a 20% tolerance will have a true ohmic value somewhere between 800 and 1200 ohms.

If you have trouble working with percentages, here is another way of computing tolerance.

$$\text{Resistance variation} = \frac{\text{rated resistance} \times \text{tolerance}}{100}$$

The answer will be the number of ohms the resistor may vary above and below its labeled value. For example,

$$\text{Resistance variation} = \frac{2000 \text{ ohms} \times 10}{100}$$

$$= \frac{20,000}{100} = 200 \text{ ohms}$$

A 2000-ohm resistor with a 10% tolerance may vary as much as 200 ohms above or below—1800 to 2200 ohms.

Resistor tolerance is not an indication of poor manufacturing. Closer tolerances can be achieved, but at greater expense. As you will discover, for a given ohmic value a 20% tolerance resistor costs less than one rated at 10%. And a 10% tolerance resistor is less expensive than one rated at 5%.

Required resistor tolerance depends on circuit design. If current flow must be controlled within close limits, the engineer specifies a 1% resistor. On the other hand, a 20% tolerance is satisfactory for circuits which have less critical operating requirements. Your radio or television set, for example, has more 20% resistors than all the other tolerances combined.

PURCHASING RESISTORS

When purchasing fixed-value resistors, describe each by its four characteristics:

1. Type (composition, wirewound, or film).
2. Value (in ohms).
3. Tolerance (5%, 10%, or 20%; 1% if wirewound or film).
4. Power rating (¼, ½, 1, or 2 watts). For higher wattages, wirewound resistors must be used.

Q6-25. Resistance limits the flow of _____ in a circuit.

Q6-26. Conductors have (low, high) resistances. Insulators have (low, high) resistances.

Q6-27. A resistor is a device that has a specific value of _____.

Q6-28. If 100 volts is applied across a 25-ohm resistor, how much current will flow?

Q6-29. A(an) _____ is used to measure resistance.

Q6-30. If 100 volts will force 2.5 amps through a device, what is the resistance of the device?

Q6-31. If 0.2 amp flows through a 50-ohm resistor, how much power (in watts) will it dissipate?

Q6-32. What is the lowest value in ohms you would expect to measure in a 5kΩ resistor having a 5% tolerance?

RESISTOR COLOR CODES

Wirewound and film resistors normally have their value in ohms and tolerance in percent stamped on them. For carbon or composition resistors a **color code** is used.

For several years, resistance values have been coded by three colored bands painted around the body of the resistor. If the tolerance is either 5% or 10%, a fourth color band is added. Position of the bands is shown in Fig. 6-12.

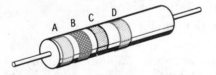

Fig. 6-12. Color bands indicate resistance value.

Colors and Numbers—Each of the colors represents one of the 10 digits—0 through 9.

Color	Number	Color	Number
Black	0	Green	5
Brown	1	Blue	6
Red	2	Violet	7
Orange	3	Gray	8
Yellow	4	White	9

The order of reading the bands is from the end of the resistor toward the middle.

The first two colors (A and B in the illustration) indicate the first two digits in the resistance value. The third band (C) indicates the number of zeros that follow the first two digits. Sometimes a fourth band (D) is present. This band indicates tolerance and will be either gold or silver. A gold band denotes 5% tolerance, silver 10%, and no fourth band, 20%. Here is an example in reading the first three bands:

Band	A	B	C
Color	Blue	Red	Orange
Numbers	6	2	3 zeros

The blue-red-orange bands signify 62 followed by three zeros and would be read as 62,000 ohms. Another example,

Band	A	B	C
Color	Violet	Green	Red
Numbers	7	5	2 zeros

Digits seven and five are to be followed by two zeros. Combined to form a number, they read 7500 ohms. Although rare, you may find a resistor with the following colors:

Band	A	B	C
Color	Violet	Green	Black

The resistance value is not 750 ohms. The third band specifies the number of zeros. Black decoded is zero. So there are no zeros after the first two digits, indicating a value of 75 ohms.

If black appears as the second color, it is read as a digit. Brown-black-red, for example, reveals that the composition resistor has a value of 1000 ohms.

Q6-33. The color bands are read from the _____ toward the _____ of a resistor.

Q6-34. The first two bands are decoded as _____.

Q6-35. The third band indicates the number of _____.

Q6-36. Decode brown-black-green.

Q6-37. Decode blue-red-red.

Q6-38. What is the color code for a 10 kΩ resistor?

Q6-39. Decode orange-green-brown-silver.

RESISTOR CONNECTORS AND CIRCUITS

There are only three different ways in which electrical or electronic parts may be connected—**series**, **parallel**, and **series-parallel**.

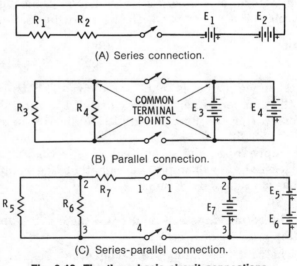

(A) Series connection.

(B) Parallel connection.

(C) Series-parallel connection.

Fig. 6-13. The three basic circuit connections.

The illustration in Fig. 6-13 shows the three different connections and also the accepted method for labeling components. R stands for resistor; E designates a voltage source. Numbers are used with the letters to identify a specific component.

Series Connection

The first figure on the opposite page shows components in series. A terminal of one component is connected to a terminal of the other. Since they are connected together in a line, R_1 is in series with R_2. Voltage sources may also be series-connected—E_1 is in series with E_2.

Parallel Connection

The second figure shows components connected in parallel. Each terminal of one component is connected to a terminal of the other. The connections are called **common terminal points**. R_3 is in parallel with R_4; E_3 is in parallel with E_4. In parallel, one component is connected **across** the other.

Series-Parallel Connection

As the third figure shows, series and parallel connections are combined to form a series-parallel arrangement. Two different combinations are illustrated. R_7 is in series with the parallel combination of R_5 and R_6. E_7 is in parallel with the series combination of E_5 and E_6.

Circuit Tracing

The method of determining the manner in which parts are connected within a circuit is called **circuit tracing**. Visualize how current would flow as you follow its path.

In the first figure on the opposite page, the same current that flows through R_1 must also flow through R_2. Current supplied by E_2 must flow through E_1 and add to the current generated by E_1. In the second figure, current is traced to one of the common terminal points. Here it must divide and flow through each **leg**, the name given to a parallel circuit path. Some of the current flows through R_3 and the rest through R_4, both currents joining again at the other terminal. Current from the E_3 and E_4 legs unites at one terminal and separates upon returning to the other terminal.

Q6-40. **In the third figure on page 126, current flows through R_7. At terminal 2 it divides and flows through both _____ of the _____ connection.**

Q6-41. **In the voltage source portion, _____ and _____ are connected in series.**

SERIES CIRCUITS

If all the components in a circuit are connected one after the other, it is called a **series circuit**. By circuit tracing, you can show that in the circuit in Fig. 6-14, the same current that leaves E_1 flows through the lamp, the ammeter, R_2, R_1, and returns to E_1 again. Therefore, the circuit must be a series type.

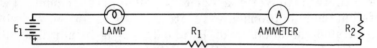

Fig. 6-14. A series circuit.

Current in a Series Circuit

It can be proved that the value of current remains the same in all parts of a series circuit by constructing the circuit on the next page. Fahnestock clips are used as terminal connections.

If the current is measured by connecting the ammeter as shown in Fig. 6-15, the reading should be between 1.6 and 1.7 milliamps. This is the value of the current entering terminal 4.

Connecting the ammeter in series with the two resistors at terminal 3, another reading may be taken. In this case the resistors are disconnected at T3 and each resistor terminal reconnected to one of the ammeter leads. Connect a short length of wire between terminals T5 and T6 to complete the circuit. The reading will again be 1.6 to 1.7 milliamps. The same results will be obtained at terminals T2 and T1.

Resistance in a Series Circuit

Total resistance in a series circuit is equal to the **sum of the resistances** of each of its parts.

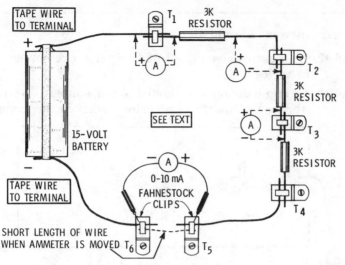

Fig. 6-15. Resistance in series.

This is logical, because the total resistance in the circuit determines the amount of current allowed to flow with a given voltage source. Therefore, to find the total resistance in a circuit, add the values of the individual resistances.

In the circuit in Fig. 6-15, the resistances are 3000 ohms each. Their sum is 9000 ohms. The ammeter also adds resistance in series. But since this resistance is normally less than 1 ohm, it adds so very little to the total that it can be disregarded.

Q6-42. In a series circuit, all of the parts are connected in _____.

Q6-43. R_1 and R_2 are connected in series. Their values are 3000 ohms and 1500 ohms, respectively. If current through R_1 is 2 milliamps, what is the value of current flowing through R_2?

Q6-44. Draw a schematic of the three resistors as they are connected in the above diagram. Show how an ohmmeter (schematic symbol) would be connected to read total resistance of the three.

Q6-45. If the ohmmeter measures 9000 ohms, how much current will flow if the three resistances are connected across a 15-volt battery?

Your Answers Should Be:

A6-42. In a series circuit, all of the parts are connected in **series.**

A6-43. Current through R_2 is also 2 milliamps. (Current through all parts of a series circuit is the same.)

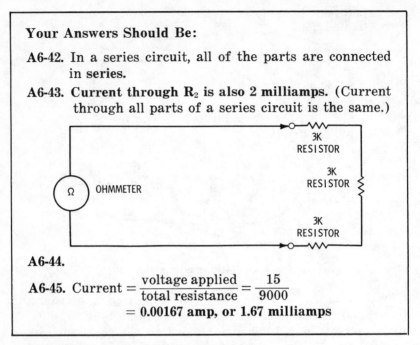

A6-44.

A6-45. Current $= \dfrac{\text{voltage applied}}{\text{total resistance}} = \dfrac{15}{9000}$
$= $ **0.00167 amp, or 1.67 milliamps**

Voltage Distribution in a Series Circuit

The voltage of a source is distributed **across and within** any load connected to it. Although this is a simple statement, the concept is often misunderstood.

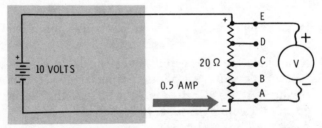

Fig. 6-16. Voltage distribution in a series circuit.

If a 20-ohm resistor is connected across a 10-volt source, as shown in the illustration in Fig. 6-16, a voltmeter reading across the resistor will be 10 volts. This means that the voltage of the source is not only applied **across** the load, but it also exists **within** it.

The **taps** (connections) shown are equal distances apart. If the lower test probe is moved to tap B, the voltmeter will be across ¾ of the resistor. And ¾ of the total voltage is 7.5 volts. Half the resistance (between C and E) will result in a measurement of 5 volts. From D to E is ¼ of the resistance and ¼ of the voltage, or 2.5 volts.

Can voltage distribution be estimated without making the measurements? Yes, and the reason is based on the familiar relationship that exists between voltage, current, and resistance and is as follows:

$$\text{Current} = \frac{\text{voltage}}{\text{resistance}}, \text{ or } I = \frac{E}{R}$$

If you do not know the value of voltage applied across a resistance of 20 ohms, but you do know the current through it is 0.5 ampere, how would you determine the voltage? You can find the value of voltage by reasoning that E/R must be a ratio that equals ½. Since R is 20, E would have to be 10 volts. Or you can restate the relationship to read **E = IR,** meaning current multiplied by resistance. To prove that it is the same equality, ½ amp times 20 ohms does equal 10 volts.

Voltage developed across a resistance is termed an **IR drop,** or, substituting E for IR, it may be called a **voltage drop.** "Drop" does not indicate voltage has been lost. Instead, it identifies the amount of voltage existing between two points of a resistance when current is flowing.

The IR (or voltage) drop between points A and E in the illustration is 10 volts. IR equals 10 volts. What is the voltage (IR drop) between taps A and B? I is still 0.5 amp, but the value of R is different. It is ¼ of the total resistance, or 5 ohms. Therefore, E = IR = 0.5 × 5 = 2.5 volts.

Q6-46. What is the value of voltage between taps A and C?

Q6-47. What is the voltage drop between taps B and E?

Q6-48. What is the IR drop between taps B and D?

Q6-49. The sum of the resistances in a series circuit is equal to the total _____ of the load.

Q6-50. The sum of the _____ _____ in a series circuit is equal to the total voltage across the load.

Examples—The schematic for three 3000-ohm resistors and a 15-volt battery circuit can be drawn to look like Fig. 6-17.

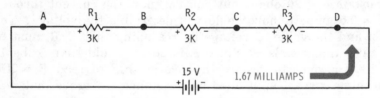

Fig. 6-17. Schematic diagram of a series circuit.

Voltage distribution principles are the same for this circuit as they were for the tapped resistor. Since the load contains three resistors of equal value, the voltage drop across each will be ⅓ of the source voltage, or 5 volts. To prove this, multiply the resistance of one of the resistors times the current in the circuit to give the voltage. This voltage will not be exactly 5 volts because 1.67 milliamps was rounded off to the next highest whole number. If you make the measurements with a voltmeter, you will find the distribution principle correct by a reading of 5 volts.

Note that each resistance is marked with polarity signs (minus and plus). The voltage across the resistor is just as real as that of the voltage source and, if the voltage is dc, the resistor has negative and positive terminals. When taking voltmeter readings, resistor polarity must be known. Circuit tracing is the best way to determine the polarity. The terminal that current enters is minus, and the one from which it leaves is plus.

Try the same reasoning on a series circuit containing resistors of unequal value.

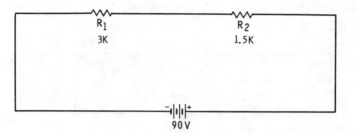

Fig. 6-18. A series circuit with unequal resistance values.

R_1 is twice the value of R_2 in the circuit in Fig. 6-18. Both are in series across 90 volts. How do you find the voltage drop across each resistor?

This can be done by either of the two methods discussed—determining proportional distribution across each resistor, or by using $E = IR$. By the proportion method it is necessary to determine what ratio (or fraction) one resistance is of the total.

$$\frac{R_1}{R\ (total)} = \frac{3000}{4500} = \frac{2}{3}$$

Two-thirds of 90 volts is 60 volts. So the drop across R_1 is 60 volts and across R_2, 30 volts.

$$I = \frac{E\ (total)}{R\ (total)} = \frac{90}{4500} = 0.02\ \text{amp}$$

Then, by using the IR relationship,

$$E = I \times R_2 = 0.02 \times 1500 = 30\ \text{volts}$$

Since the two methods are based on the same voltage distribution principle, either method provides the correct answer.

Q6-51. The (left, right) end of R_1 above is negative.

Q6-52. a. Draw a schematic of two resistances in series and supplied by 50 volts dc. R_1 is 2000 ohms and R_2 is 3000 ohms. Show all polarity marks.
   b. What is the voltage drop across R_1?

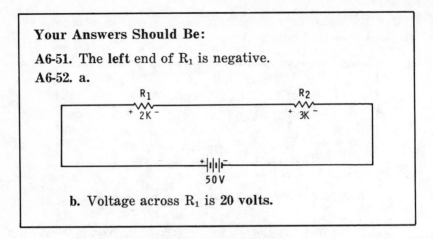
PARALLEL CIRCUITS

If all the components are connected across each other, the circuit is a **parallel circuit**. In the example shown in Fig. 6-19, the components are all connected to the same terminal (a wire in this case) and are, therefore, in parallel.

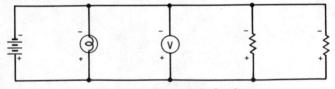

Fig. 6-19. A parallel circuit.

Polarity across each component is determined by circuit tracing. The terminal that current enters is negative.

Voltage Distribution in a Parallel Circuit

Each component (the lamp, the voltmeter, and each resistor) is connected across the voltage source. Thus, the voltage drop across each part is the same value as the source. This is true even though the resistance of each component may be different.

Current in a Parallel Circuit

Each component in a parallel circuit draws its own separate current. Each leg is connected directly to the voltage source, which means each leg can be considered as a separate circuit to determine its current.

Fig. 6-20. Two equal resistances in parallel with a voltage source.

In the diagram in Fig. 6-20, two equal resistors are shown as being in parallel across a single voltage source.

To find the current through R_1, divide the voltage across the resistor by the value of R_1. The result of this calculation is 0.04 amp. Since both resistances are equal and have the same voltage source, the current through R_2 must also be 0.04 amp. Both currents are supplied by the same voltage source, so the total current drawn must be 0.08 amp.

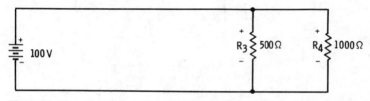

Fig. 6-21. Two unequal resistances in parallel with a voltage source.

Using the same reasoning $(I = E/R)$, it will be found that the current through R_3 in the circuit in Fig. 6-21 is 0.2 amp. The current through R_4 is 0.1 amp. The total current is 0.3 amp.

Q6-53. In a parallel circuit, voltage across each leg is (the same as, different from) the voltage at the source.

Q6-54. In a series circuit, voltage across each resistor is (the same as, different from) the source voltage.

Q6-55. In a parallel circuit, total current is the (same as, sum of) currents in each leg.

Q6-56. In a series circuit, total current is the (same as, sum of) currents in each resistance.

Q6-57. a. R_1 (20 ohms), R_2 (40 ohms), and R_3 (60 ohms) are in parallel across a 12-volt dc source. Draw the schematic.
b. Find the total current and the current in each leg.

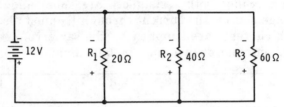
Resistance in a Parallel Circuit

How would you find the total resistance in the parallel circuit you drew in A57 above?

At this point you have used two of the three arithmetic statements that express the relationship existing between voltage, current, and resistance. To find current

$$I = \frac{E}{R}, \text{ or current} = \frac{\text{voltage}}{\text{resistance}}$$

To find voltage

$$E = IR, \text{ or voltage} = \text{current} \times \text{resistance}$$

The third way the relationship can be stated is

$$R = \frac{E}{I}, \text{ or resistance} = \frac{\text{voltage}}{\text{current}}$$

You know the total voltage across the circuit (12 volts), and you found the total current through the circuit (1.1 amps).

What is the total resistance of the circuit? Using the resistance formula above, the answer is approximately 10.9 ohms.

As you suspected, total resistance is smaller than the smallest resistance in the parallel network. Total current is the sum of the parallel currents and is, therefore, an amount that can flow only if the total resistance is smaller than that in any of the legs.

Total resistance cannot be found by adding the values of the individual resistances. The sum would be a resistance much larger than any one of the resistances. This would mean the total current would be smaller than any of the leg currents. Obviously, such a solution cannot be correct. For those who like to work with numbers, total resistance can be obtained by adding reciprocals.

$$\frac{1}{R_T} = \frac{1}{R_1} + \frac{1}{R_2} + \frac{1}{R_3} + \ldots + \frac{1}{R_n}$$

The electrical wiring in your home consists of parallel circuits. This includes the ceiling fixtures, wall outlets, and whatever else is energized electrically. Each parallel circuit is fused. If you plug one too many appliances into a circuit, the fuse blows. You have just learned the reason why. You added one more resistive path that draws current. As a consequence, total current increased beyond the capacity of the fuse, and it performed its job. Comparisons between series and parallel circuits are given in Table 6-2.

Table 6-2. Comparison Between Series and Parallel Circuits

	Series Circuit	Parallel Circuit
Voltage	Divides across resistances	Same voltage across all resistances
Current	Same current through all resistances	Divides through each resistance
Resistance Total	Sum of all the individual resistances	Less than the smallest resistance

Q6-58. Two 6-ohm resistors are in parallel across a 6-volt battery. What is the total resistance?

SERIES-PARALLEL CIRCUITS

A **series-parallel circuit** contains a combination of series-
and parallel-connected components. The simplest example is
the one shown in Fig. 6-22.

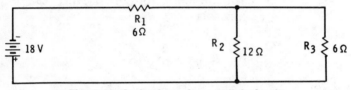

Fig. 6-22. A simple series-parallel circuit.

The best way to work with a series-parallel circuit is to
reduce all parallel combinations to an equivalent resistance.
When this is done, the total current or the total resistance
for the resulting series circuit can be readily found.

In the example shown above, how would you find the total
resistance? Think about it before you continue reading. Yes,
you could do it with reciprocals, but there is another method
that is more easily applied, even when the resistance values
are difficult.

Cover all the circuit except for the parallel network. Apply
a mythical voltage, the value of which is easily divisible by
either resistance. Perform the E/R division to find the mythical current flowing in each leg. Divide the sum of the currents
into the mythical voltage to find the **real** total resistance. The
following table uses three different voltages to show it will
work with any assumed voltage.

Mythical Voltage	6 Volts	12 Volts	24 Volts
I = E/R$_2$ (12 ohms) :	0.5 amp	1 amp	2 amps
I = E/R$_3$ (6 ohms) :	1.0 amp	2 amps	4 amps
Total I is:	1.5 amps	3 amps	6 amps
R = E/I:	**4 ohms**	**4 ohms**	**4 ohms**

The total resistance (4 ohms) is the equivalent resistance of the parallel network. The 4 ohms is in series with 6 ohms for a total circuit resistance of 10 ohms (add resistances in a series circuit). The total circuit current (E/R) is 1.8 amps.

There are many different combinations of series-parallel circuits. One that is slightly more complex is shown in Fig. 6-23. Some of the questions at the bottom of the page refer to this circuit.

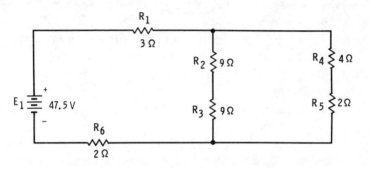

Fig. 6-23. A more complex series-parallel circuit.

Q6-59. Current in the above circuit flows through R_1 (before, after) it flows through the parallel network.

Q6-60. Total series-parallel circuit resistance is readily solved by reducing _____ resistance to an equivalent _____ resistance.

Q6-61. In a series circuit _____ divides among the resistances.

Q6-62. In a parallel circuit _____ divides among the resistances.

Q6-63. In a series circuit _____ is the same for all resistances.

Q6-64. In a parallel circuit _____ is the same for all resistances.

Q6-65. What is the total current in the above circuit?

Q6-66. What is the arithmetic statement for finding current?

Q6-67. For finding resistance?

Q6-68. For finding voltage?

Your Answers Should Be:

A6-59. Current in the above circuit flows through R_1 **after** it flows through the parallel network.

A6-60. Total series-parallel circuit resistance is readily solved by reducing **parallel** resistances to an equivalent **series** resistance.

A6-61. In a series circuit **voltage** divides among the resistances.

A6-62. In a parallel circuit **current** divides among the resistances.

A6-63. In a series circuit **current** is the same for all resistances.

A6-64. In a parallel circuit **voltage** is the same for all resistances.

A6-65. The total current for the circuit is **5 amps.** If you missed it, do it again. Guidance is on the preceding pages.

A6-66. $I = \dfrac{E}{R}$ which means: current $= \dfrac{\text{voltage}}{\text{resistance}}$

A6-67. $R = \dfrac{E}{I}$ which means: resistance $= \dfrac{\text{voltage}}{\text{current}}$

A6-68. $E = IR$ which means: voltage $=$ current $\times$ resistance.

You have now accumulated quite a bit of experience working with the current-voltage-resistance relationship that exists in all dc circuits. This same relationship also holds true for any portion of a circuit.

You were guided very closely when solving for current, voltage, or resistance, so you may not recall or be aware of the care that must be taken in applying the three different arithmetic statements. For this reason, the necessary precautions are summarized.

1. The correct algebraic forms of the arithmetic statements are

$$I = \frac{E}{R}, \; R = \frac{E}{I}, \; E = IR$$

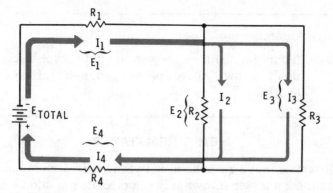

Fig. 6-24. Voltage and current distribution in a series-parallel circuit.

2. When using any one of the three formulas, I must be expressed in amperes, E in volts, and R in ohms. When values appear with milli-, kilo, mega-, or other prefixes, they must be converted to the basic units of amperes, volts, or ohms. (See conversion table 2-1.)

3. When using any one of the three formulas, values must be taken from the same portion of the circuit. Study the diagram in Fig. 6-24 for a few moments.

If you are solving for **total content,** you must use only the values that truly represent **total voltage** and **total resistance.** You cannot use E_1 (voltage across R_1) because it is not the total voltage. You cannot use the equivalent parallel resistance because it is not the total circuit resistance.

There is danger also in selecting incorrect values when seeking a solution for a portion of the circuit. If you are working with R_1, be sure the current you use is I_1 and the voltage is E_1 (volts across R_1).

Always label values to identify the circuit areas to which they belong (R_1, E_1, I_1, etc.).

OHM'S LAW

You probably have heard of or read about **Ohm's law.** Do you know what it is? Your answer should be yes. You have been working with it ($E = IR$; $I = E/R$; $R = E/I$) throughout this entire chapter.

Q6-69. In the above circuit, E_2 is 6 volts and I_1 is 120 milliamps. What is the value of R_1?

METER RESISTANCE

All meters have resistance between their terminals. When you connect a meter into or across a circuit, you add resistance to that circuit.

Ammeters

Ammeters are always connected in **series** with the circuit through which current is to be measured. As a result, the same current flows through the ammeter that flows through the circuit. The familiar connection is shown in Fig. 6-25.

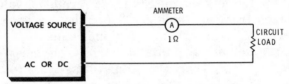

Fig. 6-25. Proper ammeter connection.

The resistance of most ammeters is less than 1 ohm. Added in series with the load resistance, very little change is made to the total resistance. If the load were 10,000 ohms, for example, the new total resistance would be 10,001 ohms—hardly enough change to make a significant difference in the current. If the load were only 1 or 2 ohms, however, a difference would be noted.

Ammeters are never connected in parallel. What would happen if an ammeter were connected as shown in Fig. 6-26?

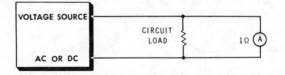

Fig. 6-26. Improper ammeter connection.

You are right. It would be damaged. But do you know why? The full voltage of the source is applied across the 1-ohm resistance of the meter. Even if the source were only a 1.5-volt dry cell, more than an ampere of current would flow through the meter. Some meters are designed to handle that amount of current, but a multimeter is not.

If an ammeter is connected to a wall outlet (117 volts ac) how much current would try to flow through the meter? More than 115 amps! However, when the current increased to 20 amps, the protective fuse in the house circuit would break the circuit. But the 20 amps would certainly damage the meter, and, under certain conditions, the person holding the meter could be injured.

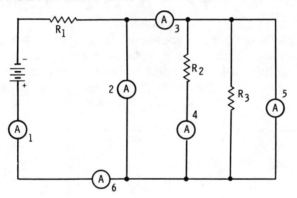

Fig. 6-27. Find the errors in this circuit.

Q6-70. There are six ammeters connected into the above circuit. Some are connected improperly. Which numbers are they?

Q6-71. A dc ammeter has a polarity that must be observed. The negative test lead must be connected to the _____ side of the circuit.

Q6-72. This means that circuit current must enter the _____ terminal of the meter and leave by the _____ terminal.

Q6-73. The (left, right) side of ammeter 3 in the above circuit is the negative terminal.

Q6-74. Current to be measured should not exceed the _____ _____ of the meter.

Voltmeters

Voltmeters are always connected in **parallel** with the component across which voltage is to be measured. As a result, the same voltage appears across the component and the meter. The illustration in Fig. 6-28 shows the proper connections.

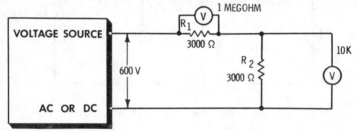

Fig. 6-28. Proper voltmeter connections.

The internal resistance of a voltmeter is normally very high. The higher it is, the greater the meter accuracy.

In the illustration, the voltmeter across R_1 has an internal resistance of 1,000,000 ohms. It will change the current flowing in the circuit very little, probably not enough to vary the true reading. As an example, assume that 0.1 amp is flowing before the voltmeter is connected; 300 volts (IR) should be read across R_1. The voltmeter adds 1,000,000 ohms in parallel, however. A little figuring shows that the equivalent resistance is now 2991 ohms, which will increase the current to 0.1001 amp, lowering the voltage reading to 299.4 volts. This is usually close enough for most purposes.

144

A voltmeter with an internal resistance of 10,000 ohms presents a different result. Using similar arithmetic, you find the new current to be 0.113 ampere and a voltage reading of 260.7 volts, almost 40 volts less than it should read.

Voltmeter Sensitivity—Internal resistance of a voltmeter is given in terms of meter sensitivity (ohms per volt). Resistance varies with the range settings. To find the internal resistance, multiply the ohms/volt rating by the maximum number of volts in a range. With a 20,000 ohms/volt meter, the following resistances are obtained:

10-volt range: 10 V × 20,000 Ω/V = 200,000 ohms
50-volt range: 50 V × 20,000 Ω/V = 1,000,000 ohms
250-volt range: 250 V × 20,000 Ω/V = 5,000,000 ohms

Corresponding resistances of a 5000 ohms/volt meter are

10-volt range: 50,000 ohms
50-volt range: 250,000 ohms
250-volt range: 1,250,000 ohms

The 20,000 ohms/volt meter is undoubtedly the more accurate meter. It will cost a few dollars more to attain this accuracy, but there will be times when you will be glad you paid extra for it.

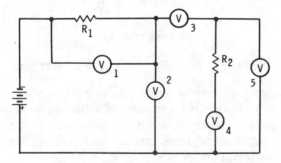

Fig. 6-29. Find the errors in this circuit.

Q6-75. In the circuit in Fig. 6-29, which voltmeters are improperly connected in the circuit?

Q6-76. The internal resistance of a voltmeter is (higher, lower) than that of an ammeter.

Q6-77. What will happen to a circuit if a voltmeter is connected in series with it?

WHAT YOU HAVE LEARNED

1. Resistance is a property of all materials which limits the flow of current.
2. Conductors have a low resistance; insulators have a high resistance.
3. Since voltage causes a certain amount of current to flow and resistance limits the amount that will flow, there is a special relationship between current, voltage, and resistance. This relationship is expressed by the following:

$$I = \frac{E}{R}, \text{ or current} = \frac{\text{voltage}}{\text{resistance}}$$

$$R = \frac{E}{I}, \text{ or resistance} = \frac{\text{voltage}}{\text{current}}$$

$$E = IR, \text{ or voltage} = \text{current} \times \text{resistance}$$

4. The unit of resistance is the ohm. The value of resistance in ohms can be measured with an ohmmeter.
5. Current flowing through a resistance generates heat. If temperature rises greatly, electrical resistance of the material increases.
6. Resistors are designated by construction (wirewound or composition) and by intended use (fixed, adjustable, or variable).
7. Resistances are rated by their heat-dissipating capability in terms of watts.
8. Resistor tolerance is given as a percentage value which indicates the amount a resistor may vary above or below the labeled value.

9. Four characteristics of a resistor must be known when purchasing a resistor. These are type, value, tolerance, and power rating.
10. Wirewound resistors have their value stamped on the body (the tolerance may also be included). Composition resistors are read by decoding colored bands painted around the body of the resistors.
11. Resistors, or any other electrical/electronic component, have only three possible ways in which they can be connected—series, parallel, and series-parallel. These terms are also the names of the circuits in which they appear.
12. Algebraic and arithmetic statements of Ohm's law are used to determine I, E, or R in a circuit or a portion of a circuit.
13. In a series circuit:
    a. Total voltage is divided among the load resistances.
    b. Current is the same through all the resistances.
    c. Total resistance is the sum of all the resistances.
14. In a parallel circuit:
    a. Source voltage appears across all the resistances.
    b. Total current divides among the resistances.
    c. Total resistance is less than the smallest resistance.
15. Never connect an ammeter in parallel with a circuit.
16. Never connect a voltmeter in series with a circuit.
17. Never connect an ohmmeter to a circuit which is connected to a voltage source.

7

Understanding Transistors

what you will learn

Transistors were invented in the early 1950s. Since that time millions have been used in a variety of electronic devices. The transistor is the basic element of the circuits used in communicating with the satellites spinning in space. In this chapter you will learn something of what a transistor is and what it can do. You will also be given details about interesting transistor circuits that you may want to build.

WHAT IS A TRANSISTOR?

The term **transistor** comes from the combination of two words—"transfer resistor." A transistor **transfers** small values of electrical energy into larger values. Weak voltages (such as those representing sound) can be made strong enough by transistor circuits to operate a speaker.

A transistor is actually a variable **resistor,** but not the ordinary type you have just studied. It is unique because its **resistance can be varied electrically.** Small incoming voltages cause small currents to flow through the transistor. The small currents make corresponding but large changes in the transistor resistance, causing other transistor currents to make equally large changes in value. The output signals are, therefore, similar to, but stronger than, the input signals.

Transistors are constructed from materials classified as **semiconductors.** Germanium and silicon are examples.

HOW TRANSISTORS ARE USED

Transistors come in several shapes and sizes. There are now hundreds of varieties, each one having special characteristics that make it different than the others for specific applications. The three types shown in the diagram are representative of the shape and size of most transistors. An **alphanumeric** (letter and number) code is assigned to each type of transistor. This designation, with the aid of a handbook, identifies the operating characteristics of each particular transistor.

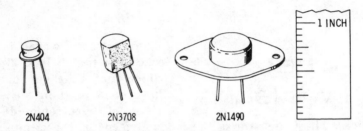

Fig. 7-1. Typical transistors.

Amplifiers

Transistors, as well as several other devices, are capable of converting small voltages or currents into larger ones. The process is called **amplification** (to enlarge). Electronic circuits that accomplish this function are known as **amplifiers**. Most of the circuits used in electronic equipment of all types are designed as amplifiers.

Vacuum tubes, for example, were used in amplifier circuits long before transistors were developed. Transistors or integrated circuits have almost replaced the tube as an amplifying device in most applications for the following reasons:

1. Due to their small size and weight equipment size is smaller.
2. Their low operating voltages allow lightweight and inexpensive power supplies to be constructed.
3. They are relatively noise free, thus permitting signals to be amplified without certain types of distortion.

Transistor amplifiers are used in radios, television receivers, tape recorders, phonographs, and a host of military, commercial, and industrial electronic equipment.

Control Circuits

In control circuits, a transistor acts as an electrically operated switch. Such circuits are built into electronic clocks, testing devices, digital computers, etc.

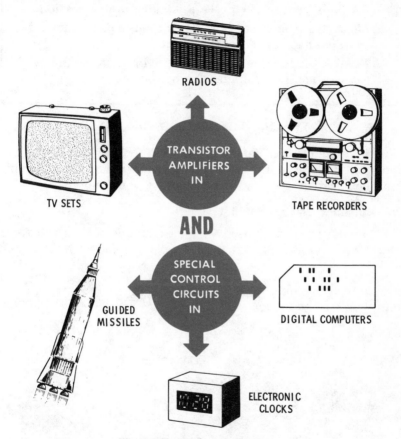

Fig. 7-2. Transistor applications.

Q7-1. A transistor is a device that can convert (small, large) signals into (small, large) signals.

Q7-2. This conversion process is called _____.

Q7-3. A transistor can amplify because its internal _____ can be electrically varied.

Q7-4. A transistor is _____ in size, _____ in weight, and can use _____ voltages for operation.

TYPICAL TRANSISTOR CIRCUITS

One type of transistor circuit will be explained to give you an understanding of transistor operation. The circuit can be constructed as an experiment if you desire.

A Transistor Voltage Divider

A transistor can be used with resistances in series to form a voltage divider. Construction details are shown in Fig. 7-3.

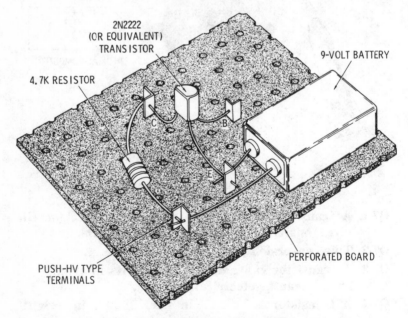

2N2222
(OR EQUIVALENT)
TRANSISTOR

9-VOLT BATTERY

4.7K RESISTOR

C

B

E

PUSH-HV TYPE
TERMINALS

PERFORATED BOARD

Fig. 7-3. A transistor voltage divider.

Be careful when handling the transistor because the fragile wires break very easily.

Two schematics of the circuit are shown in Fig. 7-4. One shows an incomplete symbol for the transistor—the correct version will be given and described later. The other schematic shows the transistor as a variable resistance. The slanted arrow indicates the resistance is variable.

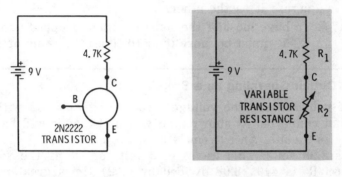

(A) Actual circuit.   (B) Equivalent circuit.

Fig. 7-4. Schematic diagrams of the transistor voltage divider.

NOTE: Until you have more experience, do not connect any ohmmeter directly across the transistor leads. Certain ranges of some ohmmeters develop enough current to destroy small transistors. The 4700-ohm resistor in the above circuit acts as a current-limiting resistance to prevent the 9-volt battery from doing the same thing.

Based on your knowledge of resistances in series, what value must R_2 be in order to obtain a voltage reading of 4.5 volts at terminal C? Yes, it must be the same value as R_1 (4700 ohms) to divide the value of source voltage equally between them.

Q7-5. What approximate value must R_2 have if nearly all the source voltage is measured across R_1?

Q7-6. To have most of the source voltage dropped across R_2, R_2 must be (more, less) than 10 times the value of R_1.

The Circuit Operating as a Switch

If you measure the voltage across the transistor with the circuit connected as shown, you will find that it will be very close to 9 volts. An extremely accurate meter might measure it as 8.98 volts. This leaves 0.02 volt for R_1. If the voltage across R_2 is 449 (8.98 divided by 0.02) times greater than the voltage across R_1, the resistance of R_2 must be the same number of times larger. This works out to be a value of over 2,000,000 ohms. The illustration in Fig. 7-5 shows how the voltmeter is connected across the transistor.

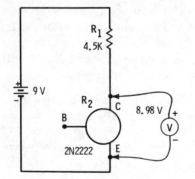

Fig. 7-5. Measuring the voltage across the transistor.

Under these conditions, current in the circuit would be very small. In effect, the high resistance of the transistor is very close to being an open circuit—no current flowing.

The structure of the transistor material will cause this resistance to change when current flows from B to E.

By adding a large resistor (80,000 ohms) from the positive side of the battery to the B terminal of the transistor, approximately 2 milliamperes will flow through the transistor. The meter reading, as shown in Fig. 7-6, is near 0 volts.

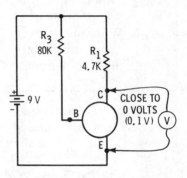

Fig. 7-6. Voltage across the transistor with a negative voltage applied to terminal B.

In effect, the path through the transistor from C to E has decreased from a very high resistance to a very low resistance. The transistor is operating as if it were a switch. The pair of diagrams below show the transistor acting as a switch. With no current through B, the transistor acts as an open switch; with current it acts as a closed switch.

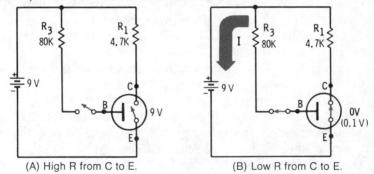

(A) High R from C to E. (B) Low R from C to E.

Fig. 7-7. A transistor can act like a switch.

Q7-7. When a small amount of current enters the B terminal, the transistor acts as a(an) _____ switch and almost (all, none) of the source voltage appears across it.

Q7-8. With no current through B, the C to E switch is (open, closed) and almost (all, none) of the source voltage can be measured across the terminals.

In practice, the transistor switch is opened and closed by electrical signals applied to the B terminal. These will be discussed later in the chapter.

TRANSISTOR SYMBOLS AND CONNECTORS

Transistors have their own symbols and methods of being connected within a circuit, the same as other electrical and electronic components.

Symbols

There are two types of transistors—npn and pnp. The difference between the two is the type of material used in their construction and, as a result, the direction that current flows between terminals.

Symbols for both types are shown in Fig. 7-8. The 2N2222 (used in the experiment) is an npn transistor.

The only difference in the two symbols is the direction of the arrowhead. Be able to recognize either one. Current not only flows through the two types in different directions, but they are connected into a circuit differently.

Also note the names for the B, C, and E terminals. **Base, collector,** and **emitter** identify the significant parts of a transistor. The letters B, C, and E are the notations often used in schematics.

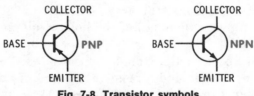

Fig. 7-8. Transistor symbols.

Shown in Fig. 7-9 are the basic circuits for npn and pnp transistors and the flow directions for base (I_B), collector (I_C), and emitter (I_E) currents. Study them carefully.

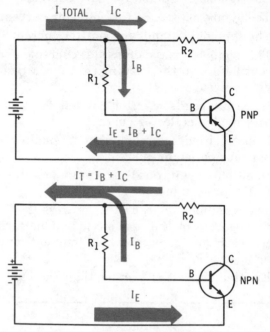

Fig. 7-9. Basic transistor circuits.

Q7-9. The symbol I_B stands for _____ current.

Q7-10. The symbol I_C stands for _____ current.

Q7-11. The symbol I_E stands for _____ current.

Q7-12. The symbol I_T stands for _____ current.

Q7-13. In both circuits, emitter current is equal to _____ current plus _____ current.

Q7-14. In both circuits, total current is equal to _____ current plus _____ current.

Q7-15. In either circuit, _____ current and _____ current are the same value.

Q7-16. The terminals of R_1 and R_2 nearest the transistor are negative in the (npn, pnp) circuit.

7Q7-17. Emitter current is (greater, less) than the base current.

A SIMPLE CONTROL CIRCUIT

A control circuit can be made from the transistor voltage divider used in the first part of this chapter. The control circuit will be used as a means of applying the principles you have learned about transistors. Enough details are furnished so you can construct and use the circuit if you desire.

Circuit Construction

The control circuit is very similar to the other transistor circuits that have been discussed. A schematic diagram (including parts detail) is shown in Fig. 7-10.

Resistors R_1 and R_2 connect the transistor base to the positive terminal of the 9-volt battery. Resistor R_1 can be any low resistance material or device. Its purpose is to interrupt the base circuit when the event to be detected takes place.

Resistor R_3 connects the collector to the positive terminal of the battery. The negative battery terminal is connected to the emitter. A lamp is connected in parallel with the transistor and will light when the event is detected.

The battery is not shown as a symbol in the schematic. Arrowheads, however, point to the proper terminal connections. This technique is used to save space and reduce clutter in the schematic diagram.

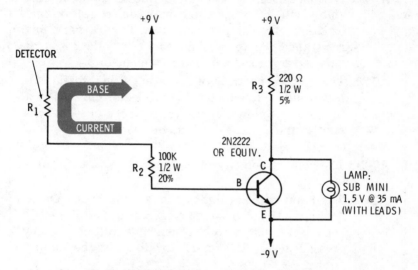

Fig. 7-10. A simple alarm circuit and detector.

Q7-18. When base current flows through the transistor, a very (high, low) resistance appears between collector and emitter.

Q7-19. When resistance between C and E is high (nine, zero) volts will be impressed between collector and emitter.

Q7-20. The voltage polarity of R_3 is negative on the (collector, battery) side.

Q7-21. The voltage polarity of R_2 is positive on the (base, detector) side.

Q7-22. The lamp (will, will not) light as long as base current is flowing.

How the Circuit Operates

The preceding questions established the fundamental principles of transistor operation. Now these principles will be used to explain how the circuit operates.

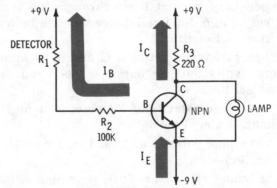

Fig. 7-11. Current flow in the transistor alarm circuit.

When detector R_1 remains unbroken, the following conditions exist:

1. Base current flows through the transistor.
2. The transistor acts as a closed switch.
3. Zero voltage appears between C and E.
4. The lamp does not light.

When the detector is broken (opened):

1. Base current stops flowing.
2. The transistor acts as an open switch (from C to E).
3. Voltage appears between C and E.
4. Voltage is applied across the lamp, causing it to light.

The Detector—A simple detector can be made from a strip of aluminum foil. Taped to a door or a window, it will break when either is opened, thus lighting the alarm. Construction details are shown in Fig. 7-12.

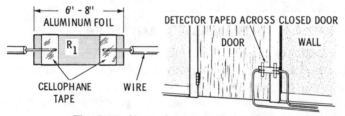

Fig. 7-12. Alarm detector construction.

Q7-23. The lamp is out when _____ current is flowing.

Q7-24. The lamp is on when C-to-E resistance is (high, low).

Q7-25. When the lamp is lit, there is (minimum, maximum) voltage across R_3.

Q7-26. When the detector breaks, there is (zero, maximum) voltage across R_2.

Q7-27. The value of base current is equal to _____ current minus _____ current.

Q7-28. In a circuit using an npn transistor, base current flows (toward, away from) the source, and emitter current flows (toward, away from) the transistor.

Q7-29. (Collector, emitter, base) current is equal to total current flowing through the source.

A TRANSISTOR AMPLIFIER

A transistor can be used to amplify voltages. As you recall, amplify means to increase amplitude or value. In other words, a weak signal (radio wave or audio voltage, for example) can be made stronger by passing it through an amplifying circuit.

Alternating Current or Voltage

Before you begin thinking about amplifiers you should become more familiar with alternating voltage. You know that an alternating current reverses itself periodically. During one period of time, current flows in one direction. During the next period, current flows in the opposite direction. The flow of current does not make the change instantaneously; it does not move in one direction, then pause before it moves in the other. The changes occur over a period of time, regardless of how short the time may be.

The illustration in Fig. 7-13 shows a graph of how an ac voltage (or current) changes direction. The same picture can be seen on an electronic test instrument, called an **oscilloscope.**

The dimensions of the graph show voltage values vertically and elapsed time in seconds horizontally.

You will note that this ac **sine wave** (as it is called) takes 2 seconds to change from zero volts to its maximum of 2 volts. It rises rapidly during the first second (to approximately 0.7 of its maximum value). It then rises less and less rapidly until its full value is finally reached at the end of the remaining second.

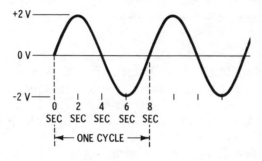

Fig. 7-13. An ac sine wave.

From zero to 2 seconds the voltage is rising in the positive direction. It decreases to zero volts, following the same shape curve, in the next 2 seconds. At this point it has completed one half of its full cycle. This portion of the **waveform** is called the **positive half cycle.**

During the **negative half cycle,** it repeats the first half, except that now it moves in the negative (or opposite) direction. From zero volts, it **increases** in value until it reaches the maximum negative voltage—minus 2 volts. This takes 2 seconds. In the next 2 seconds, the voltage **decreases** from its maximum negative voltage back to zero.

The rise and fall of the positive and negative half cycles are identical. The only difference is the direction—one is from zero to a maximum positive voltage and back to zero, while the other is from zero to a maximum negative voltage and back to zero. The ac voltage in your home follows the same pattern. It completes 60 full cycles every second (60 positive half cycles and 60 negative half cycles).

Q7-30. A weak signal can be increased in amplitude by passing it through a(an) _____ circuit.

Q7-31. An ac cycle consists of two half cycles; one is _____ and the other is _____.

Gain

The behavior of an amplifier is described in terms of **gain.** Without going into lengthy detail, amplifiers are designed to operate at some definite voltage level during periods when no signals are applied. In the illustration in Fig. 7-14, the voltage level of the input side is −2 volts. The output side operates at +90 volts.

If an ac signal that is changing from +1 volt to −1 volt is applied to the input, the signal voltage will add to the dc voltage already present at the input. For example, when the signal is maximum positive, the +1 volt is added to the −2 volt dc level to produce a −1-volt input to the amplifier at that time. When the signal swings in the negative direction, −1 volt and −2 volts become −3 volts. In other words, the swing of the input is from −1 to −3 volts, or a **change in voltage** of 2 volts.

Assume that the characteristics of the amplifier are capable of making the corresponding output voltage changes shown in the diagram with such an input signal. The operating level of the amplifier output is +90 volts. During the first half cycle of the input, the output changes from +90 volts to +80 volts. During the second half cycle, the swing is from +90 volts to +100 volts.

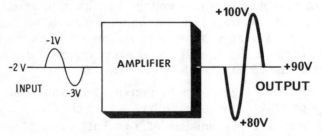

Fig. 7-14. An amplifier operates at certain voltage levels.

This means that an **input voltage change** of 2 volts caused an **output voltage change** of 20 volts (from +80 to +100). The output change was 10 times that of the input, so the **gain** of this amplifier is 10.

How would you express this in arithmetic form? Like this:

$$\text{Gain} = \frac{\text{change in output voltage}}{\text{change in input voltage}}$$

Or, if you want to find the change in output for a given amplifier with a particular input signal:

Change in output E = gain × change in input E

Problem—If a certain signal causes a change of 3 volts on the input side of an amplifier and a corresponding change of 48 volts at the output, what is the gain?

$$\text{Gain} = \frac{\text{output change (volts)}}{\text{input change (volts)}} = \frac{48}{3} = 16$$

Problem—Input voltage of an amplifier changes from −1.5 volts to −4.5 volts. The amplifier gain is 12.3. What is the voltage change in the output?

$$
\begin{aligned}
\text{Output change (E)} &= \text{gain} \times \text{input change (E)} \\
&= 12.3 \times 3 \text{ volts} \\
&= 36.9 \text{ volts}
\end{aligned}
$$

Amplifiers can be designed with gains ranging from very small to very large. The amount of gain desired is first determined and then an amplifier is selected that has such a gain.

Q7-32. An alternating voltage appears at the input of an amplifier. It causes the input voltage to vary from −2 volts to −1.2 volts. What is the change in input voltage?

Q7-33. The same change in input voltage produces an output waveform that swings from +12 volts to +42.6 volts. What is the gain of the amplifier?

Q7-34. An amplifier has a gain of 39; the input voltage swings from −5.25 to −5.05 volts. What will be the change in the output voltage?

Q7-35. Would a steady dc voltage applied to the input of an amplifier cause the circuit to amplify?

A Simple Transistor Amplifier

You may wish to construct the transistor amplifier shown in the schematic diagram in Fig. 7-15.

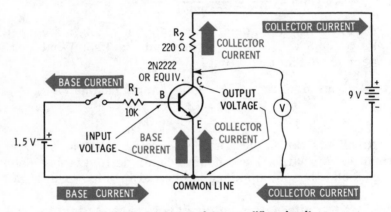

Fig. 7-15. A simple transistor amplifier circuit.

A change in input voltage will occur if the switch in the line connected to the base is opened and closed. The change can be observed on a voltmeter with its leads connected to the base and the common line. Output voltage changes can also be read with a voltmeter connected as shown. Fig. 7-16 shows the voltage changes that will occur.

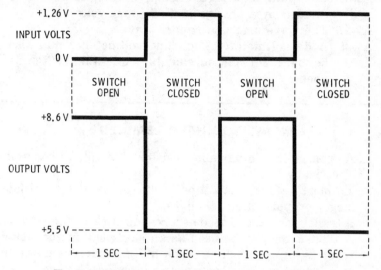

Fig. 7-16. Voltage changes in the transistor amplifier.

With the switch open, the input registers 0 volts and the output +8.6 volts. With the switch closed, the input changes to +1.26 volts and the output to +5.5 volts. The same readings occur each time the switch is opened and closed.

If the switch were opened and closed at one second intervals, as shown in the graph, the cycles would repeat themselves at a steady rate. The **frequency** would be one cycle every two seconds as opposed to the household electrical ac frequency of 60 hertz. (Hertz is one unit of frequency equal to one cycle per second and is abbreviated Hz.) Because they have straight sides and flat tops, the waveforms in the graph are called **square waves.**

Q7-36. What is the gain of the transistor amplifier circuit just described?

Q7-37. The output waveform rises rapidly, remains at a steady value, and then decreases rapidly to its original level. What type of waveform is it?

Q7-38. What type of voltmeter would you use to measure the output when the switch is closed?

WHAT YOU HAVE LEARNED

1. A transistor is a variable resistor that can be controlled electrically.
2. Transistors use small input voltage changes to produce larger output voltage changes.
3. A transistor may be used as a resistor in a voltage divider.
4. A transistor may be used as an electrical switch. It will react as an open switch when no base current flows and as a closed switch when base current does flow.
5. There are two types of transistors, npn and pnp. In an npn, base and collector currents flow away from the transistor; emitter current flows toward the transistor. In a pnp transistor, the base and collector currents flow toward the transistor; while emitter current flows away from the transistor.
6. The ac voltage rises to, and falls from, its maximum voltage periodically. The polarity alternates between positive and negative directions.
7. The gain of an amplifier is determined by dividing the change in input voltage into the corresponding change in output voltage.

8

How To Solder

You will now learn how to make permanent electrical connections using a procedure called soldering. You will be shown how to work with the necessary tools and hardware. The fundamentals are easily learned. Skill in soldering requires careful practice. When you complete this chapter you will be able to apply the fundamentals of soldering, select and use the proper tools, properly prepare an iron for soldering, make proper mechanical and electrical connections, and make and disconnect soldered joints.

THE PURPOSE OF SOLDERING

Solder is a special metal mixture applied to electrical connections to prevent **oxidation**. **Soldering** is the process of applying the right amount of the mixture to join two or more pieces of metal.

Oxidation is the result of a chemical reaction between air and certain metals. When iron oxidizes, rust forms on its surface. When copper oxidizes, a dull, insulating film forms on its surface. The film has a high resistance to current flow. In order to retain a good electrical connection, the film must be removed and prevented from forming again.

To protect iron from rusting, it is painted. To protect copper and similar metals from oxidation, they are coated with solder. In addition to providing a good electrical path, the solder also adds to the mechanical strength of the joint when two wires are joined.

THE PROCESS OF SOLDERING

Solder is applied to a connection with heat. A tool that supplies this heat is a **soldering iron** or a **soldering gun.**

Solder

Solder is a metal alloy (mixture) containing tin and lead. The alloy most usually used in electronic work is 60/40 solder —60% tin and 40% lead.

Other tin/lead ratios are available but are not recommended for use in electronic equipment or small electrical appliances. 50/50, 40/60, and 30/70 are all solders that have less tin content than the 60% desired. Tin is the metal in the alloy that leaves the bright, tightly bound, conductive coating produced by soldering.

When tin and lead are combined, the resulting alloy has a lower melting point than either of the metals by itself. Tin heavy ratios, such as 70/30, 80/20, and others are available, and probably would permit a better soldering job, but the melting point of these combinations is too high for most work. Higher heat is required for application. High temperatures damage or destroy components—resistors, coils, and transistors, for example—and excessive heat chars, burns, or melts insulation on wires and components. 60/40 solder seems to be the best balance between the need for a properly soldered joint and the maximum heat that can be tolerated.

Solder comes wrapped on spools or supplied in coil form. Its shape is usually round and wire-like. However, it also can be produced in flat, ribbon-like lengths.

An oxidation preventer called **rosin** must be used when soldering in electrical or electronic work. Although a connection may be cleaned until it is bright and shiny (as it should be), application of heat during the soldering process causes the connection to rapidly oxidize again. Rosin melts at low temperatures and forms a coating to protect the metal from air while heat builds up to melt the solder.

Rosin can be purchased separately in a tube or a can, but solder that contains a core of rosin can be purchased. Such a combination is called **rosin-core solder** and is recommended for all soldering in electrical circuits.

Some solders contain an **acid paste** in the core. **Do not use**

acid-core solder. It corrodes metals used in electrical and electronic equipment. The corrosive effect eats away metal and leaves a nonconducting layer having a very high insulation resistance.

Soldering Tools

A soldering iron or gun is used to melt solder. For most electrical work, either is usually suitable. However, each has certain advantages for specific jobs. The gun (pistol-shaped) provides heat within a few seconds at the touch of a trigger, but is usually heavier, larger, and more difficult to use in close quarters than an iron of the same rating. An iron reaches soldering temperature slowly and must be unplugged to cool, but is usually less expensive than a gun.

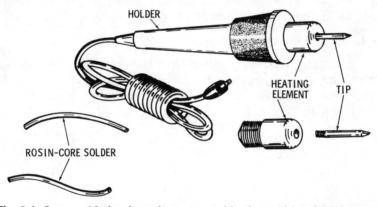

Fig. 8-1. Some soldering irons have removable tips and heating elements.

The heating power of an iron and gun is rated in watts. Irons are available with ratings from 6 watts or less, to 500 watts and more. The larger irons are for heavy industrial use. Irons rated from 10 to 75 watts are best for most electronic work. Remember to use no more heat than necessary to obtain a properly soldered connection.

A variety of tips can be used with the iron pictured in Fig. 8-1. At least one should be a **spade tip** which is most useful for general-purpose soldering.

Q8-1. For electronic work, solder should be a(an) _____/
_____ tin/lead alloy and have a _____ core. The soldering iron should be _____ _____ _____ watts.

SOLDERED CONNECTIONS

You will often hear that a good, tight **mechanical connection** is required before soldering. This means a connection that will remain tightly bound between wire and wire, or wire and terminal. It must not be movable during the soldering process and it must be strong enough to resist jarring loose under normal equipment operation.

Tools for Making Connections

In addition to an iron, other tools for soldering are needed. Although primarily used when making a mechanical connection, these tools have other useful purposes. Three such tools are shown in Fig. 8-2.

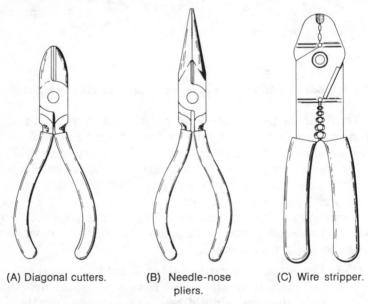

(A) Diagonal cutters.   (B) Needle-nose pliers.   (C) Wire stripper.

Fig. 8-2. Necessary tools for electrical or electronic work.

Diagonal Cutters—Diagonal cutters, also called "dikes," are used for cutting wire, trimming leads and terminals to length, and stripping insulation from wire.

Long-nose, or needle-nose, pliers—Long-nose pliers are used to hold materials in place, form wire to the shape of terminal connections, make wire splices, and as a means of diverting soldering heat from delicate parts.

Metal Stand (Rest) for Iron—A soldering-iron stand is a useful accessory for soldering work. One can be purchased, but you can easily make one from a small tin can. Flatten one side of the can by bending a small area of the top and bottom rims outward. Just enough bending is needed to keep the can from rolling. A dent placed in the top forms a seat for the iron. Placed on the rest, the iron will not be free to burn other material on the bench, and contact with the can helps draw away excess heat.

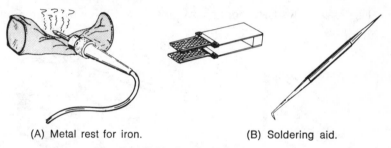

(A) Metal rest for iron.   (B) Soldering aid.

Fig. 8-3. Soldering accessories.

Soldering Aids—Soldering aids come in several shapes. They are used primarily to remove excess solder during soldering and unsoldering. The one shown in Fig. 8-3 has a sharp point and hook. An ice pick can also be used.

Q8-2. "Dikes" can be used to cut wire and remove _____.

Q8-3. Long-nose pliers càn be used as a means of bypassing _____ during a soldering job.

Q8-4. An ice pick can be used to remove excess _____ while unsoldering.

Q8-5. A wire splice must be mechanically tight so the wires will not _____ during soldering.

Q8-6. _____ must be removed from bare wires before they can be soldered.

Connections Between Terminals and Wires

Fig. 8-4 shows the proper way to splice a wire to a terminal.

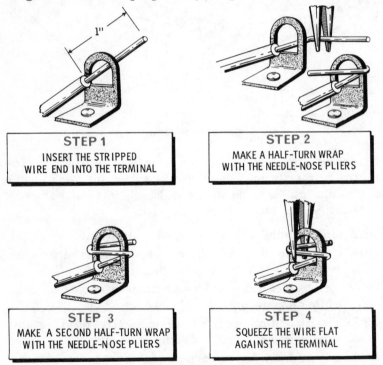

STEP 1
INSERT THE STRIPPED
WIRE END INTO THE TERMINAL

STEP 2
MAKE A HALF-TURN WRAP
WITH THE NEEDLE-NOSE PLIERS

STEP 3
MAKE A SECOND HALF-TURN WRAP
WITH THE NEEDLE-NOSE PLIERS

STEP 4
SQUEEZE THE WIRE FLAT
AGAINST THE TERMINAL

Fig. 8-4. Connecting a wire to a terminal.

Preparing the Wire—Remove from 9 to 25 millimeters (⅜ to 1 inch) of insulation from the end of the wire, depending on the type of connection to be made. If the wire is not bright and shiny, it probably has an oxidized film on it. The film should be removed by scraping with a knife or by using a lead cleaning tool or fine sandpaper.

Preparing the Terminal—A new terminal (one never used) will be ready for the splice. It has been coated with metal to which solder will readily adhere. If the terminal is not shiny and silvery in color, it should be cleaned with sandpaper, a lead cleaning tool, or a knife point. If the terminal has been previously used, remove all excess solder. This can be done by heating the terminal and wiping the melted solder away with a rag or a small wire brush. Care must be taken to prevent any of the molten solder from being splattered onto your skin or into your eyes, or from shorting other connections.

Making the Connection—Follow the steps shown in the illustration in Fig. 8-4 to make a connection. If two or more wires are to be secured to a terminal, make the connection for each in the same manner. This can be done individually or by the wires together as a group.

After making the connection, test it carefully. Check the length of the bare wire left between terminal and insulation. Too much bare wire (over 9 mm) may be the cause of future shorts. After soldering, wiggle and tug on the connection. Make sure the wire does not move on the terminal.

Q8-7. From _____ to _____ mm or more of bare wire should be exposed for connection to a terminal.

Q8-8. The wire should be wrapped through and around the terminal with _____-_____ pliers.

Q8-9. Wire should be squeezed tight on the terminal for a strong _____ connection.

Q8-10. Wires and terminals must be clean. They can be cleaned with a knife or _____.

Q8-11. Exposed wire should be no longer than _____ mm from terminal to insulation.

Q8-12. If the connection moves while being wiggled or tugged, what should be done?

Wire-to-Wire Connections

Wire-to-wire connections are also made by splices that are mechanically strong. Three common splices are shown below.

Pigtail Splice—This splice is easily made. Cross the wires, as shown in Fig. 8-5, and begin twisting the wires together. The twist should be started by hand and completed with pliers to make sure the splice is tight. Do not exert too much pressure or the wires may break at the bottom of the splice.

After the splice has been soldered, fold it back and alongside one of the wires. Wrap electrical insulating tape around the splice. The wrap should have two or three layers if the wire is used for 117-volt purposes.

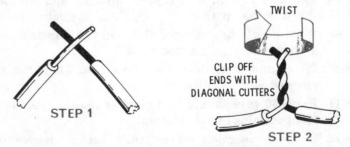

Fig. 8-5. The pigtail splice.

Western Union Splice—Developed in the early days of the telegraph, the Western Union splice is neat and mechanically tight if properly made. Starting from the crossed-wire position, wrap one wire neatly around the other. Keep the coils close together. Do the final tightening with a pair of pliers. Then straighten the wires as in Step 3. Wrap the second wire closely around the first and tighten. There should be at least four coils on each side of the junction. Solder, then wrap the splice with electrical tape (Fig. 8-6).

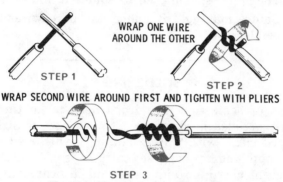

Fig. 8-6. The Western Union splice.

Stranded-to-Solid Splice—Twist the stranded wire into a straight, tight spiral and wrap around the solid wire as in Step 2 of the Western Union splice. Make at least a half-dozen turns. Fold and tightly crimp the end of the solid wire over the turns. Solder and wrap the splice with tape.

Q8-13. A pigtail splice is made by _____ wires together.

Q8-14. A Western Union splice is made by _____ one wire around the other to form tight coils.

Q8-15. In a stranded-to-solid splice, the _____ wire should be wrapped around the _____ wire.

Q8-16. After a splice has been formed by hand it should be tightened with _____.

Q8-17. The splice should then be _____ and _____.

Q8-18. Before making any splice, the bare wires should be _____.

SOLDERING

The tips of some soldering irons need to be **tinned.** Any tip that is corroded needs to be retinned and, if dirty, needs to be wiped clean with a rag. Tinning (with solder) is the process of applying a protective coating of solder on the copper tip to prevent corrosion and as an aid to heat transfer. The process is shown in Fig. 8-7.

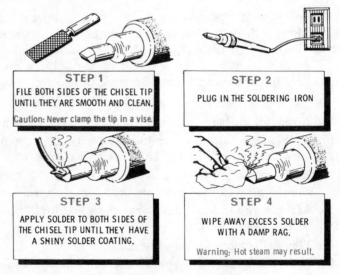

Fig. 8-7. Tinning a soldering iron.

Soldering to Printed-Circuit Boards

The areas where the leads of mounted components come through a printed-circuit board are called **lands**. Parts are mounted from the reverse side and leads drawn through on the foil side. The leads should be pulled up fairly tight so the components lie flat on printed-circuit board as shown in Fig. 8-8. Then, bend the leads down against the board, along the exposed copper and clip them off just beyond the bend, leaving about 9 mm against the land. The small bends in the leads hold the component in place while it is being soldered.

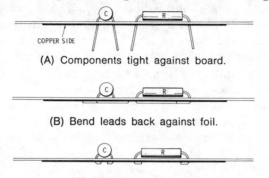

COPPER SIDE

(A) Components tight against board.

(B) Bend leads back against foil.

(C) Clip off excess lead.

Fig. 8-8. Attaching components to a printed-circuit board.

Because the connections are physically small it does not take much heat to bring them to soldering temperature. Use a small pencil iron with a rating between 10 and 30 watts. Printed-circuit board soldering is a little more precise than for standard connections. There must be enough heat for good solder wetting, but not so much that there is danger of the copper foil being lifted from the board.

The control of solder flow is easier if a smaller-diameter solder is used. With the smaller connection, and the quicker transfer of heat, a shorter time is required to solder the connection.

It is most important to use only enough solder to fill the crevices between the wire and the land area of the foil (Fig. 8-9). Insulated spaces between circuits are frequently very small. You must not allow a bridge of solder to form across two circuit sections. If it does, use one of the desoldering methods described below.

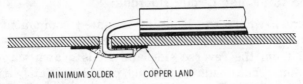

MINIMUM SOLDER   COPPER LAND

Fig. 8-9. A soldered connection on printed-circuit board.

Desoldering Connections

Repairs on printed-circuit boards and other terminals are much easier to make if the solder on the connection is removed. A steel-bristle brush may be used on printed-circuit boards to brush the hot solder away when melted by an iron; this has its dangers in that small flecks of solder can get into the circuit and cause shorts. Another method is to turn the board upside down, and let the solder wick down into a hot iron. The fastest, easiest, and most complete desoldering method is by means of a vacuum system that sucks the solder away from the terminal and completely off the board.

The desoldering irons work as follows: a bulb is depressed,

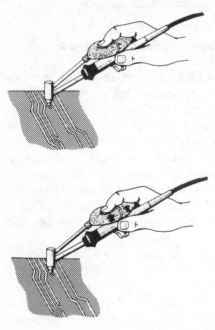

(A) Depress bulb, then apply iron to connection.

(B) Release bulb.

Fig. 8-10. Desoldering with desoldering iron.

SOLDER
(USE SPARINGLY)

TOUCH THE SOLDER TO
THE FRONT OF THE LUG.

TOUCH THE IRON TO
THE BACK OF THE LUG

Fig. 8-11. Soldering to a terminal.

the iron is applied to the connection, and when the solder has melted, the bulb is released and the vacuum draws the solder up a tiny tube in the tip of the iron (Fig. 8-10). The iron is lifted away, placed over a container or waste area, and the bulb depressed again to eject the solder in the tube.

Soldering a Wire to a Terminal

A good solder joint has just enough solder to bond the splice; has a smooth semishiny appearance; and has edges that seem to blend into the terminal and wire (Fig. 8-11). Any other appearance is a sign of improper soldering.

The iron should be applied to the back side of the terminal and held firmly against the wire. After a few seconds, touch the solder to the wire at the front of the terminal. If the solder readily melts, the connection is sufficiently heated. If the connection is not hot enough, remove the solder and continue heating until it is. Do not touch the solder to the iron.

When the connection is hot enough, touch the solder to the wire in front of the terminal. Rosin will melt and coat the connection. Let the solder flow to the terminal and the iron. It will flow properly when the connection is the right temperature. When the connection has been coated (not just covered) with solder, remove the length of solder. Leave the iron on the connection to boil away the rosin. (If any rosin remains under the solder, an insulation barrier will be formed.) Now remove the iron and let the connection cool. **Do not move the joint while it is cooling.**

Q8-19. The copper tip of a soldering iron must be _____ prevent corrosion.

Q8-20. Solder is applied to a connection (at the same time as, sometime after) the iron is applied.

Q8-21. Solder is applied to the (connection, iron).

Soldering a Splice

Quite often heat from the soldering iron may melt or burn the insulation on a wire being soldered, or may damage a com-

Fig. 8-12. A heat sink.

ponent (resistor, transistor, etc.). Heat from the iron travels rapidly down the wire.

To prevent this, use the long-nose pliers as a **heat sink.** Grip the wire between the terminal and insulation, or the lead between the equipment and connection, as shown in Fig. 8-12. The iron in the pliers dissipates most of the heat before it can travel further down the wire.

The correct positions of iron, connection, and solder while soldering splices is shown in Fig. 8-13. **Let the solder flow to the iron through the connection.**

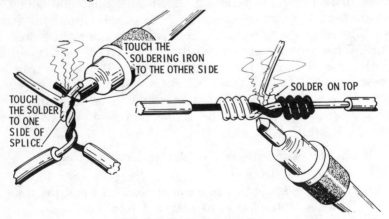

Fig. 8-13. Soldering a splice.

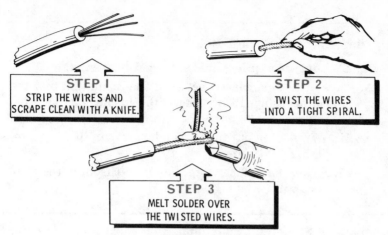

Fig. 8-14. Tinning stranded wire.

Tinning Stranded Wires—Conductors made up of several small wires are called **stranded wires.** When the insulation is stripped back, the small wires tend to fan out. In fact, the wires are very difficult to keep together, especially when an attempt is made to fasten them under a screw-down terminal. Connecting stranded wire to a terminal often leaves an untidy job with one or more of the small wires pointing away from the connection. Many shorts have been caused by stray wires from a stranded conductor. This problem can be solved by tinning the strands (Fig. 8-14).

After the tinned strands have cooled, shape the wire to fit the terminal to which it will be connected. If it is a screw-down terminal, place the end of the wire in the jaws of the long-nose pliers and bend the wire over the rounded back to form a loop. Slip this around the screw. If the wire is to go into a terminal or around a solid wire instead, use the pliers to form it to the object.

Q8-22. _____ _____ make a good heat sink.

Q8-23. A heat sink placed between the insulation and the _____ will help keep the insulation from being melted or burned.

Q8-24. Wires of a stranded conductor can be kept together by _____.

Q8-25. Solder should flow to the _____ through the _____.

Checking a Soldered Connection

One way to test a solder job is with an ohmmeter. A good solder connection has zero resistance. The procedure is shown in Fig. 8-15. In making the test, touch the test probes to the metal of the pieces that are joined. Do not touch the probes to the solder itself.

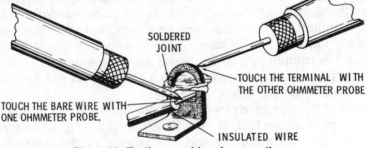

SOLDERED JOINT

TOUCH THE TERMINAL WITH THE OTHER OHMMETER PROBE

TOUCH THE BARE WIRE WITH ONE OHMMETER PROBE.

INSULATED WIRE

Fig. 8-15. Testing a soldered connection.

An ohmmeter test does not always reveal a poor soldering job. But if the meter reads zero and the solder looks neat, smooth, semishiny, and well blended into the metal; and you observed all the precautions, the connection should be electrically good. The precautions are restated below.

1. Never solder with a cold iron.
2. Never solder with an untinned or dirty iron tip.
3. Let the solder flow to the iron through the connection.
4. Let just enough solder flow to coat the connection.
5. Burn away all rosin. Any left beneath the solder will cause the insulating effect of a **rosin joint.**

Unsoldering Connections

It is quite often necessary to disconnect a component or a wire from a soldered terminal. The procedure is relatively simple if you take care to do it correctly and make sure that you do not burn or char any of the nearby parts. Follow the steps as outlined in Fig. 8-16.

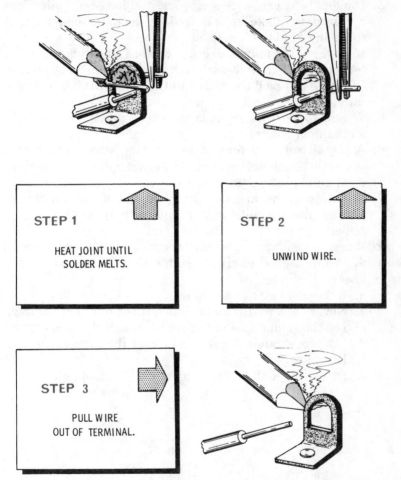

Fig. 8-16. Unsoldering connections.

Place an ice pick or soldering aid into the eye of the terminal following Step 3. This will prevent the opening in the terminal from remaining filled with solder.

WHAT YOU HAVE LEARNED

1. Rosin-core, 60/40 solder should be used in electrical and electronic work.
2. A soldering iron with a rating of 10 to 75 watts should be used for electrical and electronic work.
3. Diagonal cutters, wire strippers, and long-nose pliers are required to make splices and as help in soldering. A soldering aid is also helpful.
4. If wires or terminals are not cleaned of oxidation prior to soldering, a high-resistance connection may result. Cleaning can be done with a knife, lead cleaning tool, and sandpaper.
5. Wires joined to terminals or wires must form a tight mechanical connection.
6. A pigtail splice is formed by twisting wires together; a Western Union splice is made by wrapping each wire in tight coils around the other.
7. Soldering irons are tinned by filing off the oxidation, coating the tip with solder, and wiping off the excess solder.
8. When soldering, certain precautions must be followed to obtain a good electrical connection. These are as follows:
    a. Never solder with a cold iron.
    b. Never solder with an untinned tip or one that is dirty.
    c. Let the solder flow to the iron through the connection.
    d. Let just enough solder flow to coat the connection.
    e. Burn away all rosin.
    f. Do not move the connection while it is cooling.

9

Understanding
Transformers

what you
will learn Man has learned how
to improve the usefulness
of ac voltage by convert-
ing to higher and lower
values. In this chapter
you will learn how a transformer can accomplish this.
When you have finished, you will be able to explain what
transformers are, how they are used, and how they are
connected into circuits.

WHAT IS A TRANSFORMER?

A transformer is an electrical device which converts ac
voltages and current from one value to another. Transformers
are made in a number of varieties and sizes. Large transformers are used to furnish 117 volts ac for homes. The
voltage at the generating plant may be several thousand volts,
which is reduced to the 117-volt level by a series of transformers along the power line leading to the user. The final
step-down in voltage is usually accomplished by a transformer
on a utility pole near the user's home.

There are transformers in most homes also. Door bells or
chimes usually operate on 12 or 16 volts ac. A transformer
changes the house voltage of 117 volts to the bell-ringing
voltage. Most radios, television receivers, record players,
stereo systems, etc., contain one or more transformers. Some
of these convert the 117 volts to lower or higher voltages to
operate the sets; other transformers are used as connecting
links between circuits.

HOW DO TRANSFORMERS WORK?

Transformers contain coils of insulated wire wound on an iron frame. As you learned in earlier chapters, ac flowing through a coil develops a magnetic field that expands and contracts in step with the changes in the current. The magnetic field of one coil **induces** current to flow in the other coil by cutting through the turns of wire.

Transformer Windings

The basic transformer is constructed with two coils wound around a single **core** (iron frame). The coils are called **windings**. The input side is the **primary winding** and the output side, the **secondary winding.**

The Primary Winding—The primary winding is the input to the transformer. It receives ac voltage and current from a source. The primary of the bell transformer, for example, is connected to a 117-volt line.

The Secondary Winding—The secondary winding is the output from the transformer. Its voltage and current values are different from those in the primary. In the bell transformer (Fig. 9-1), the 117 volts applied to the primary is converted to a 16-volt ac output in the secondary.

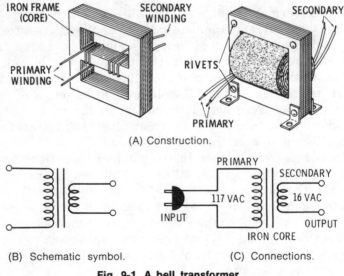

(A) Construction.

(B) Schematic symbol.   (C) Connections.

Fig. 9-1. A bell transformer.

Fundamental Principle

The transfer of energy that takes place between the coils of a transformer is called **transformer action.** Transformer action is based on the fundamental electrical principle of a **moving** magnetic field being able to induce current in a conductor.

1. **A current flowing in a conductor develops a magnetic field about the conductor.** As shown in Fig. 9-2, the direction of the lines of force in the field depends on the direction of current flow. In part A, the lines of force are counterclockwise—in part B, clockwise.

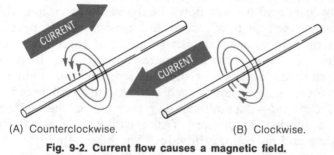

(A) Counterclockwise. (B) Clockwise.

Fig. 9-2. Current flow causes a magnetic field.

2. **Magnetic lines of force cutting through a conductor cause current to flow in that conductor.** The field must be **moving.** In a single conductor, the current is very small. If the conductor is formed into a coil, many turns will be cut by the moving field, thus developing a larger current. An example is shown in Fig. 9-3.

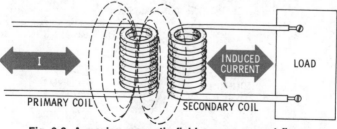

Fig. 9-3. A moving magnetic field causes current flow.

Q9-1. Magnetic lines must be (moving, stationary) to induce current in a conductor.

Transformer Action

The requirements for induced current are that magnetic lines of force must cut through a conductor and the magnetic field must be moving (expanding outward or contracting inward).

DC Current—Direct current, as you know, maintains a steady level and always flows in the same direction. Does dc induce current to flow in another conductor? It produces a magnetic field whose strength (number of force lines) is proportional to the number of amperes flowing. But the magnetic field remains steady, neither expanding nor contracting. Therefore, dc does not induce current in another conductor.

AC Current—Does alternating current induce electrons to flow in another conductor? Yes, because ac is constantly increasing and decreasing in value. The magnetic lines of force generated by the ac increase and decrease correspondingly. The magnetic field expands outward and contracts inward as the value of current changes. This means that the magnetic lines of force change direction as the current changes from the positive half cycle to the negative half cycle.

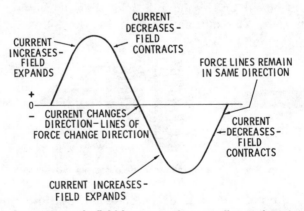

Fig. 9-4. An ac magnetic field is constantly expanding and contracting.

The diagram in Fig. 9-4 demonstrates how the magnetic field expands and contracts with the rise and fall of current. The field is in constant motion. An alternating current, therefore, induces current to flow in another conductor or coil. In this case, the induced current will also be alternating.

Energy Transfer—An applied ac voltage causes current to flow in the primary winding of a transformer. This causes a changing magnetic field which induces a current to flow in the secondary. The induced current will develop an ac voltage across the secondary winding. Therefore, it is the nature of the voltage and current in the primary to transfer energy to the secondary in the form of a voltage and current. Transformer action for one full cycle of ac voltage is shown in Fig. 9-5. Notice that the output is of opposite polarity (a phase shift of 180°).

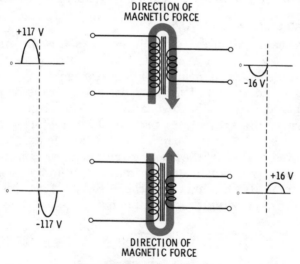

Fig. 9-5. Transformer action.

Q9-2. An ac current develops a changing _____ _____.

Q9-3. A changing magnetic field develops a(an) _____ _____ in a conductor.

Q9-4. To induce current, a field must _____ through a conductor.

Q9-5. _____ but not _____ induces current in a conductor.

TRANSFORMER CHARACTERISTICS

Now that you understand the fundamental principles of the transfer of energy (voltage and current) from primary to secondary, you are ready to learn how transformers are rated.

Basic Transformer Circuit

A schematic diagram of a basic transformer circuit is shown in Fig. 9-6. This circuit demonstrates the principles and characteristics of nearly all transformers.

If you decide to build the circuit, a bell transformer can be purchased in most hardware stores. The circuit contains a fuse (note the symbol) to protect the transformer. If the secondary of the transformer should accidentally have a short placed across it, the short circuit will be reflected back into the primary, causing the primary current to increase to a large value. If this happens, the fuse will blow instead of the transformer windings burning out.

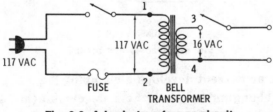

Fig. 9-6. A basic transformer circuit.

The lines between the primary and secondary windings indicate an iron core which provides an easier path for the magnetic field through the coils.

Voltage Ratio

One of the specifications for rating transformers is stated in terms of a **voltage ratio**. This ratio is a comparison of primary voltage to secondary voltage, and is written as

$$\text{Voltage ratio} = \frac{\text{primary voltage}}{\text{secondary voltage}}$$

Remember, the primary voltage is on the input side of the transformer and secondary voltage on the output side.

Step-Down Transformers—A **step-down transformer** is one having an input (primary) voltage larger than its output (secondary) voltage. The bell transformer is an example of a step-down transformer. Its voltage ratio is 117 to 16. It can be written as 117/16 or 117:16.

Step-Up Transformers—The input voltage of a **step-up transformer** is smaller than its output voltage. The transformer steps up the primary voltage to a higher value in the secondary. The distinction between step-up and step-down transformers is one of use only. As the diagram in Fig. 9-7 shows, the same transformer can be used for either purpose.

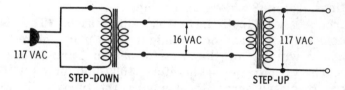

117 VAC   16 VAC   117 VAC

STEP-DOWN   STEP-UP

Fig. 9-7. Step-down and step-up connections.

Q9-6. What is the voltage ratio of the step-up transformer in the diagram?

Q9-7. A(an) _____ _____ in a transformer helps direct the magnetic field through the coils.

Q9-8. For a given magnetic field (more, less) current is induced in a straight wire than if it were wound into a coil.

Q9-9. A dc current does not induce current in a coil because its magnetic field is (moving, stationary).

Q9-10. Induced current flows in the (primary, secondary).

Turns Ratio

Since transformers must have a variety of different voltage ratios, what is there about transformer action that permits this to occur? Look at Answer 8 above and then answer the question. If one coil turn (loop) will induce a certain voltage, two turns will develop twice as much, and 100 turns 100 times as much.

Therefore, the voltage ratio between the primary and secondary windings depends on the turns ratio between the two windings. Fig. 9-8 shows an example.

Parts A and B both have 1000 turns each. (This is an example only—a transformer might have many more.) The secondary winding of the transformer in Part A has 100 turns. The turns ratio is, therefore, 1000/100, or 10/1. If 10 volts were placed across the primary, the turns ratio would produce 1 volt in the secondary. If 20 volts were applied to the primary (providing the wire could handle the increased current), the output would be 2 volts, etc.

In Part B, a turns ratio of 1000/500 (or 2/1) permits a voltage ratio of 10/5. If the primary voltage were reduced to 5 volts, there would be 2.5 volts on the secondary.

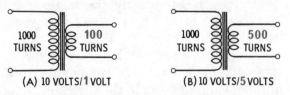

1000 TURNS	100 TURNS
(A) 10 VOLTS/1 VOLT	

1000 TURNS	500 TURNS
(B) 10 VOLTS/5 VOLTS	

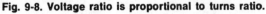

Fig. 9-8. Voltage ratio is proportional to turns ratio.

If current is doubled in the primary, the magnetic field strength will also double. Twice as many lines of force will cut the secondary and induce twice as much current. Secondary voltage will also be doubled.

But would the proportions of the voltage ratio be changed? No. To double the primary current, the primary voltage must be doubled. The voltage ratio would be increased in number but remain the same in proportion.

The reason voltage ratios are given in voltage figures instead of reduced fractions is to advise the user what the correct input voltage should be. Wire size of the windings is selected for the amount of current that will flow at that voltage. If voltage is increased beyond the rated figure, the increased current may burn the winding.

While the voltage ratio is usually given in voltage figures, the turns ratio is reduced to its lowest terms. For example, a turns ratio of 25,000/10,000 would be expressed as 5/2.

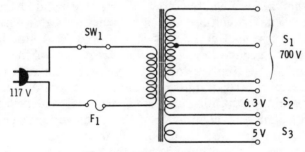

Fig. 9-9. Power transformer.

The diagram in Fig. 9-9 shows a power transformer similar to those used in some tv receivers. It has three secondary windings—S_1, S_2, and S_3. Disregard the center tap on S_1.

Q9-11. The voltage ratio of the primary to S_1 is _____.

Q9-12. The turns ratio of the primary to S_3 is _____.

Q9-13. The transformer (does, does not) have an iron core.

Q9-14. The symbol designated by F_1 is a(an) _____.

Q9-15. Would S_1 increase to 1400 volts if the primary were connected to a 230-volt source?

Frequency Rating

Another transformer rating is the ac frequency for which the transformer is designed. Frequency, as you recall, is measured in hertz. Because its voltage does not vary, dc has a zero frequency. The ac voltage varies because its value rises and falls during its positive half cycle followed by a similar rise and fall in the negative direction. The frequency of the voltage is the number of times a complete cycle repeats in a second.

Transformers are designed to operate at one specific frequency. Wire, insulation, and core material are selected to operate efficiently at the number of times the voltage (current) values rise and fall and change direction.

Reactance—You are aware that the atomic structure of a resistor or wire offers a resistance to the flow of electrons (current). Electrons find it twice as difficult to flow through a 2000-ohm resistor as through one of 1000 ohms.

Constantly changing ac current encounters a similar **reaction** when flowing in a coil. Expanding and contracting lines of force cut through the primary coil (the conductor in which they were developed) as well as the secondary winding.

As the illustration in Fig. 9-10 shows, the magnetic field induces a current in its own coil that tends to oppose the coil current. These two currents react against each other. This characteristic is called **inductive reactance.** It opposes or limits the flow of ac just as resistance limits ac or dc in a resistor.

For purposes of simplicity, only one segment of the total force lines is shown in the diagram. Keep in mind that the magnetic field actually surrounds the conductor at every point along its length.

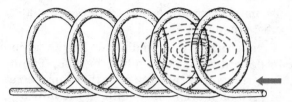

Fig. 9-10. Magnetic lines cutting an adjacent turn.

Reactance is directly related to frequency. The amount of reactance in a coil is determined by the frequency of the current and by the number of turns of wire in the coil. The greater the number of times the magnetic field changes direction in a second, the more times adjacent turns will be cut, and the greater will be the opposing current.

Coils and transformers are designed to operate at the reactance established by the designated frequency. For example, a coil may have a reactance of 30 ohms to a current whose frequency is 60 hertz. If the coil is connected to a 600-Hz source, its reactance will increase to 300 ohms. Since reactance is in opposition to ac, the current through the coil will be less with the 600-Hz source than with the 60-Hz source.

Suppose a 400-Hz transformer is connected to a 60-Hz 117-volt wall outlet. What will happen to the transformer? The reactance will be reduced to almost one seventh its 400-hertz value and almost seven times as much current will flow. The excess current will probably burn the winding. Remember to check the frequency rating before connecting a transformer to a voltage source.

Q9-16. A full ac cycle contains _____ and _____ half cycles.

Q9-17. Transformers are designed to operate at one specific _____.

Q9-18. A transformer is designed to operate at 60 hertz. What will happen if it is connected to a dc source?

WHAT YOU HAVE LEARNED

1. Transformers are electrical devices which convert voltage and current from one value to another.
2. Transformers contain at least one primary and one secondary winding. The windings are coils and are sometimes wound on iron cores.
3. Current flowing in a conductor develops a magnetic field about the conductor. Magnetic lines of force cutting through a conductor cause current to flow.
4. Transformers can be designed for ac but not for dc.
5. Transformer action is a transfer of energy. The ac in the primary generates a magnetic field which induces current in the secondary.
6. Transformers are rated as follows:

   a. Voltage ratio. The voltage ratio specifies the number of volts transferred between the two windings. The transformer can be used as either a step-up or a step-down unit, depending on which winding is used as the input.

   b. Turns ratio. A ratio of primary turns to secondary turns.

   c. Frequency. Because of ac reactance, transformers are designed for use at a specific frequency. Use at any other frequency may damage the windings.

10

Understanding Capacitors

A capacitor is another very basic but highly useful circuit component. Since it can regulate current, as do resistors and coils, the capacitor is used for this purpose in most electronic and many electrical circuits. Upon completing this chapter, you will understand what capacitors are, how they are used and connected in circuits.

WHAT IS A CAPACITOR?

A capacitor has the ability to store electrical energy. Because it can do this, it is able to control the amount and manner in which current will flow in a circuit.

Most electronic circuits consist of a combination of only three components—resistors, inductors, and capacitors. Each reacts in a different way to ac and dc voltage and current.

A **resistor,** as you recall, **controls** electricity by limiting the flow of current. This reaction is called **resistance.** It affects the flow of either ac or dc current.

An **inductor** controls electricity by regulating the flow of ac current. The magnetic field in an inductor cuts its own coils, developing a voltage that opposes a change in current. This is called **inductance.**

A **capacitor** controls electricity by also regulating the flow of ac current. It stores an electrical **charge** which opposes any change in current. This property is called **capacitance.**

HOW DOES A CAPACITOR WORK?

A capacitor, sometimes called a **condenser,** is manufactured in several shapes and sizes. A number of capacitors are shown in Fig. 10-1. You may recognize a few.

(A) Ceramic disc.

(B) Ceramic adjustable.

(C) Mica.

(D) Paper tubular.

(E) Electrolytic tubular.

(F) Paper "bath tub."

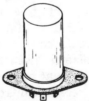

(G) Electrolytic can.

(H) Variable.

Fig. 10-1. Types of capacitors.

Several of each kind of capacitor are probably in your home. They can be found in radios, television receivers, intercoms, audio systems, and other electronic equipment. A capacitor is used with some electrical motors and even in the ignition systems of automobiles.

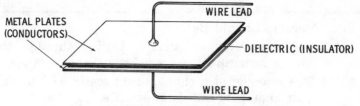

Fig. 10-2. Basic capacitor construction.

Basic Construction

Every capacitor is constructed in the same basic manner. An insulating material, called a **dielectric,** is sandwiched between two conductors (usually a pair of metal plates). A wire is connected to each plate to form the leads or terminals of the capacitor. Details are shown in Fig. 10-2.

This basic principle is elaborated upon to produce the shapes shown on the opposite page. For example, the plates of the variable capacitor are curved and exposed. Since air is an insulator, it forms the dielectric.

The tubular capacitor, as shown in Fig. 10-3, uses lengths of metal foil as the plates, which are separated by strips of treated paper to form a dielectric. Wire leads are connected to the exposed ends of the foil and the assembly is rolled into a tight spiral and placed in a case.

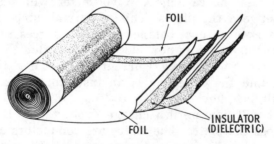

Fig. 10-3. Tubular capacitor.

Q10-1. The opposition to current flowing through the atomic structure of material is called _____.

Q10-2. The reaction of a changing magnetic field to current change is called _____.

Q10-3. The reaction of a stored electrical charge to current change is called _____.

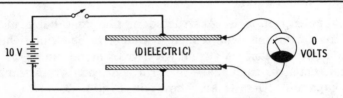

Fig. 10-4. Capacitor connected to battery.

Electrical Principle

The structure of a capacitor obeys the fundamental principles of voltage and current as applied to conductors and insulators. Assume, as shown in Fig. 10-4, that the plates of a capacitor are connected to a battery through a switch. An edge view of the plates is illustrated.

The open switch prevents the battery voltages from being applied across the capacitor. A voltmeter will show a zero voltage between the plates. This is normal, since all matter tends to seek a natural balance when no forces are applied.

Suppose the switch is now closed. A first thought might be that no current would flow. Current does not flow through an insulator, and the dielectric is an insulator. Current in the circuit will flow, however (Fig. 10-5).

Current will flow until a charge (voltage) of 10 volts appears across the plates. The plates are conductors and, therefore, have electrons free to flow as current. Electrons have a

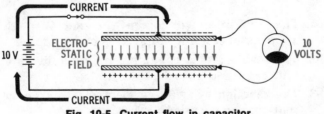

Fig. 10-5. Current flow in capacitor.

202

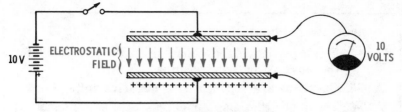

Fig. 10-6. Charged capacitor.

negative charge. They are repelled (caused to move) by the negative pole of the battery and attracted by the positive pole. The positive terminal pulls electrons away from the bottom plate of the capacitor and the negative terminal forces them to accumulate on the top plate.

A deficiency of electrons results in a positive **potential** (voltage) on the bottom plate and an excess of electrons makes the upper plate negative. Current flows, rapidly at first, but more slowly as the voltage across the capacitor builds up to the same potential as the battery. When current ceases to flow, 10 volts will be across the capacitor.

A field of force, equal to 10 volts, now exists between the two plates. This field is called an **electrostatic** force and has a direction as shown by the arrows in Fig. 10-6—from negative to positive. The excess electrons are attracted to the positive plate (whence they came). It is this attraction that develops the force. Suppose the switch is now opened. Will the electrostatic force of 10 volts disappear?

The voltage across the capacitor is still 10 volts, just as it was before the switch was opened. The excess electrons remain because there is no path for them to return to the positive plate (assuming the voltmeter has infinite internal resistance through which no electrons can travel).

Q10-4. Electrons leave the capacitor plate connected to the _____ battery terminal.

Q10-5. Electrons are repelled by a _____ voltage.

Q10-6. The plate that has a(an) (excess, deficiency) of electrons has a negative charge.

Q10-7. A(an) _____ force is set up between the plates of a charged capacitor.

Q10-8. The insulating material through which the force lines extend is called a(an) _____.

Capacitor Charge and Discharge

"Charging" is the term used when a capacitor is acquiring a potential. In the example in Fig. 10-6, the capacitor was charged to 10 volts.

Some capacitors can be charged to extremely high voltages and will retain this charge for long periods. The capacitor in the high-voltage section of a tv receiver builds up to 10,000 volts or more. So be careful when working around capacitors. A capacitor can be discharged either through normal operation of a circuit or by shorting the capacitor leads.

When the capacitor discharges, the excess electrons return to the positive plate, the difference in potential (voltage) between the two plates becomes zero, and the electrostatic force disappears. As a matter of fact, the voltmeter constitutes a circuit between the plates of the capacitor, thus forming a path for the electrons to return to the positive plate. The high resistance of the meter, however, limits the current, resulting in a long discharge time. This can be seen by watching the meter pointer (Fig. 10-7).

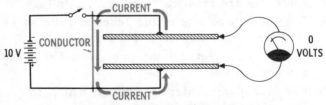

Fig. 10-7. Capacitor discharging.

Effect on DC Current—As you have seen, a capacitor **blocks** the passage of dc current. Current flows only long enough to build up a charge equal to the source potential.

Effect on AC Current—You will often hear or read that ac **flows through a capacitor.** This is not true. As long as the dielectric retains its insulating quality (the applied voltage does not become great enough to puncture a path through the dielectric) very few electrons will pass through.

With ac voltage applied, electrons accumulate first on one plate and then the other, as the voltage changes polarity. But this electron current does not change in phase (in step) with the voltage. The diagram in Fig. 10-8 shows the schematic symbol and letter designation for a capacitor, and a graph of the current and voltage relationships.

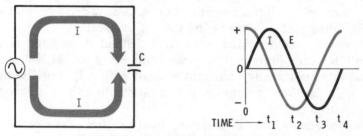

Fig. 10-8. Voltage and current in capacitor.

At time zero, the applied voltage starts to go positive. At that instant, current flow from one capacitor plate to the other is maximum. As the source voltage increases, I decreases because the charge on the capacitor is getting closer and closer to the applied E. When E reaches maximum positive (at time t_1), the capacitor is charged to the same value. Current is zero. When source E decreases toward zero volts, capacitor E is greater and causes current to flow in the opposite direction. At zero source volts, current has become maximum negative (time t_2). The difference and equality of the source voltage and the capacitor charge continue in the same time sequence for the next half cycle. As the graph shows, current is always a quarter of a cycle ahead of the source voltage.

Q10-9. Current leads ac voltage by a(an) _____ cycle in a capacitor circuit.

CAPACITOR CHARACTERISTICS

Before circuits are shown that prove the effects of a capacitor on ac and dc current, you must learn a little more about capacitor characteristics.

Units of Measurement

A capacitor is measured in terms of its capacitance, which is a definition of how many excess electrons it can store on one plate to develop a specific charge. A **farad** is the unit of capacitance just as ohm is the unit of resistance.

However, when a farad was first defined it was too large a unit for any practical purpose. Capacitors are either measured in microfarads (one-millionth of a farad), abbreviated μF, or in picofarads (one-millionth of a millionth of a farad), abbreviated pF.

Capacitance (farads) is determined by three factors:

1. Area of the plates. Larger area, greater capacitance.
2. Distance between the plates. The closer the plates, the greater the capacitance (Fig. 10-9).
3. Dielectric constant (type of material). A higher constant, a larger capacitance. The constant for air is given as 1. Paraffin paper is 3.5; mica, 6; flint glass, 9.9. For example, a mica capacitor would have 6 times as much capacitance as a capacitor with air as a dielectric, all other things being equal.

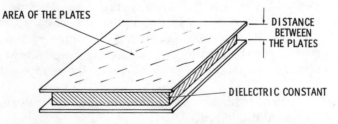

AREA OF THE PLATES

DISTANCE BETWEEN THE PLATES

DIELECTRIC CONSTANT

Fig. 10-9. Factors affecting capacitance.

Voltage Rating

If the voltage applied across a capacitor is too large, the dielectric fails to maintain its insulating qualities. It breaks down under the stress of the electrostatic force and allows current to flow from one plate to the other.

Capacitors are given a **working-voltage** rating. This rating is the highest voltage that a capacitor can withstand without the possibility of creating a short-circuit through the dielectric. The type of material and thickness of the dielectric determine what the working voltage will be.

Since the distance between plates is one of the factors which determines the capacitance and working voltage, a capacitor having both a large capacitance and a high voltage rating will also have a large plate area.

A working voltage rating pertains to a dc voltage or the peak value of ac. A more practical value of ac voltage and current is the rms value (rms stands for root mean square). The rms value is the actual "working value" of a voltage or current and is equivalent to the dc value that would accomplish the same amount of work.

The current and voltage value most often used is rms. In fact, the standard household voltage of 117 volts is the rms value. In all sine waves the rms value is equal to 0.707 of the peak value. Conversely, the peak value is equal to 1.41 times the rms value. The peak voltage of 117 volts ac is actually 165 volts.

A capacitor with a 150-volt rating will work well on 117 volts dc but not 117 volts ac. Standard practice is to use a capacitor with a working voltage about 50% higher than any voltage expected in the circuit.

Q10-10. A (thick, thin) dielectric allows more capacitance.

Q10-11. A small plate area develops (greater, less) capacitance than a larger area.

Q10-12. Glass is a (better, poorer) insulator than mica.

Q10-13. Working voltage is equal to the dc value or to the _____ value of ac.

Q10-14. A higher working voltage will be possible with a (thicker, thinner) dielectric.

Your Answers Should Be:

A10-10. A **thin** dielectric gives more capacitance.

A10-11. A small plate area will develop **less** capacitance than a large area.

A10-12. Glass is a **better** insulator than mica.

A10-13. Working voltage is equal to the dc value or to the **peak** value of ac.

A10-14. A higher working voltage will be possible with a **thicker** dielectric.

A TIMING CIRCUIT

A capacitor and a resistor can be placed in a circuit to operate as a timing device. If the values of R and C are carefully selected, the circuit can determine when an exact number of seconds has elapsed. This circuit contains components from previous circuits—a 15-volt battery, a 1-megohm resistor, a dpdt switch, and a 20-μF electrolytic capacitor (connected as shown in Fig. 10-10).

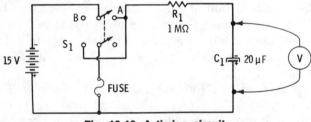

Fig. 10-10. A timing circuit.

The resistor limits the amount of current flow which charges C_1. The values of R_1 and C_1 determine the charge time.

If you have constructed the circuit, you can see the buildup of the capacitor charge (voltage) by watching the voltmeter. At the instant S_1 is moved to position A, the meter pointer begins to move quickly across the scale. As the capacitor increases its charge (in opposition to battery voltage) current starts to decrease. The pointer moves slower and slower. Before it reaches 15 volts (full charge), its movement is difficult to detect.

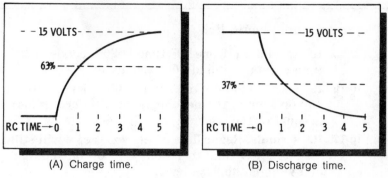

| (A) Charge time. | (B) Discharge time. |

Fig. 10-11. RC time constant.

In an RC (resistance-capacitance) circuit, charge and discharge times are measured in RC seconds. Part A in Fig. 10-11 shows the rise of voltage across a capacitor during charge.

RC is a quantity obtained by multiplying R (ohms) by C (farads). Any capacitor in an RC circuit (such as this one) charges to 63% (actually 63.2%) of its final value (battery voltage) in one RC second. In 5 RC seconds it reaches full value. The arithmetic statements are

$$R(ohms) \times C(farads) = time(seconds) ; \text{ or}$$
$$R(megohms) \times C(microfarads) = time(seconds)$$

Since R is one megohm and C is 20 microfarads in the circuit on the opposite page, the capacitor charges to 63% of its full charge in 20 seconds. Sixty-three percent of 15 volts is 9.45 volts. When the meter pointer reaches this value you know that 20 seconds have passed since moving S_1 to position A.

If you move S_1 to position B, the capacitor discharges at the rate shown in Part B (the exact reverse of Part A). The capacitor discharges to 37% of its full charge in one RC second, a value of 5.55 volts. The **RC time constant,** as it is called, holds true for any voltage.

Q10-15. How long does it take the capacitor in the circuit on the opposite page to charge to 15 volts?

Q10-16. If R_1 were 10 megohms and C_1 were 16 microfarads, how long would it take C_1 to charge to 9.45 volts?

Q10-17. If a 50-volt battery were used, how many seconds would it take the capacitor to reach 63% of full charge? What would the voltage be at that time?

DC BLOCKING

By alternately manipulating the two switches in the preceding circuit, you can simulate what the capacitor will do if ac voltage is applied. Closing one switch charges (increasing ac voltage) the capacitor, and closing the other switch (and opening the first switch at the same time) discharges (decreasing ac) the capacitor.

Whether dc or ac, current in the circuit did not pass through the capacitor. Instead, it flowed back and forth through the circuit, collecting first on one plate and then on the other. The effect is that of ac being passed through the capacitor. This characteristic of a capacitor is used in most electronic circuits where it is necessary to control alternating current, and still allow it to flow through the circuit.

Transistors and vacuum tubes operate in electronic circuits with dc voltage applied to their elements. In nearly all such circuits, an ac signal is applied to the input of the circuit (to be amplified, for example). The ac must be allowed to pass through the amplifier, but the dc voltage must not. A capacitor can be used for this purpose—**blocking dc.** The circuit in Fig. 10-12 the next page demonstrates how this is accomplished.

DC Blocking Circuit

The circuit in Fig. 10-12 uses a 15-volt battery, a 3000-ohm resistor, and a 20-μF electrolytic capacitor.

How many volts would you expect to read with the meter probes connected to points A and C? Fifteen volts, the full battery voltage, is correct. Points A and C are directly connected to the battery. How many volts will appear between points B and C?

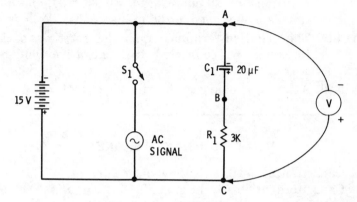

Fig. 10-12. A dc blocking circuit.

To develop a voltage across R_1, current must flow through it. At the instant the circuit is connected, current flows from the lower plate of the capacitor through R_1 and the battery to the upper plate. In 5 RC times, the capacitor is fully charged and current stops flowing.

Applying the voltmeter across R_1, after this time gives a reading of zero volts. When current is not flowing through a resistor, a voltage drop is not present.

Applying an AC Signal—Closing S_1 applies an ac signal (voltage and current) across the same load. The ac current flows back and forth from one plate of the capacitor to the other through the signal source. It also develops a voltage across R_1 that corresponds to the changes of the ac signal. Connections from B and C to another circuit would apply the changing voltage to the input of the second circuit. And no dc would interfere with its proper operation.

Q10-18. In Fig. 10-12 (without ac applied), how long does it take C_1 to charge to 63% of 15 volts?

Q10-19. The _____ terminal of an electrolytic capacitor must be connected to a negative voltage source.

Q10-20. After C_1 is charged, _____ will not flow through R_1.

WHAT YOU HAVE LEARNED

1. Capacitance controls the flow of ac current.
2. Capacitors are used in many electrical and electronic circuits.
3. A capacitor is a pair (or pairs) of plates separated by a dielectric. The dielectric, an insulator, does not pass current as long as the applied voltage is kept within the capacitor rating.
4. When a capacitor is connected to a voltage source, electrons move from one plate through the circuit and accumulate on the other plate. This charges the capacitor electrically and develops an electrostatic field between the plates. A capacitor discharges when the excess electrons on one plate return to the other plate.
5. Direct current is blocked by a capacitor. Alternating current appears to pass through by alternately charging and discharging the capacitor.
6. The effect of an electrostatic charge on ac is to cause current in a circuit to follow the ac wave pattern a quarter of a cycle in advance of the voltage.
7. A capacitor is measured in farads, microfarads (μF), and picofarads (pF). It has a working-voltage rating that should not be exceeded.
8. Capacitors can be used in a timing circuit where charge and discharge time is measured in terms of an RC time constant. Another use for a capacitor is to block dc from those parts of a circuit where it is not desired.

Understanding Diodes

In the early days of electricity, a terminal for an electrical device was called an electrode—a battery electrode, for example. When a device having two electrically active terminals was developed, it was termed di(two) -ode(electrode). A diode is a device through which current passes readily in one direction but with great difficulty in the other. In this chapter you will learn of the two different families of diodes, what they are, how they work, and what useful purposes they serve in a circuit.

WHAT IS A DIODE?

A **diode** is an electrically operated device which has two elements (or terminals). If a voltage source is applied to these elements in the correct polarity, current flows through the diode. However, if the polarity is reversed, very little (if any) current passes through.

There are two general families of diodes—**vacuum tube** and **solid state**. A **vacuum-tube diode** is constructed with the two elements enclosed in a glass or metal envelope. Current flows from one element to the other through a vacuum or, in some units, through a special type of gas.

A **solid-state diode** contains two dissimilar metals or two different types of semiconductor materials. Current flows between the junction formed by the metals or materials.

HOW DO DIODES WORK?

Although both types of diodes conduct current in one direction only, the electrical principles which permit them to do this are different.

Their appearances are different also. The largest of the devices is the vacuum-tube diode; the others shown in Fig. 11-1 are representative shapes of the solid-state variety.

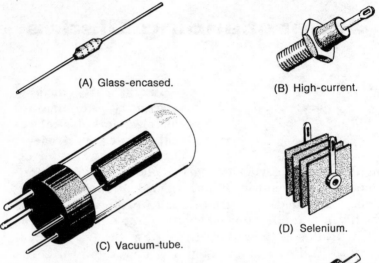

(A) Glass-encased.

(B) High-current.

(C) Vacuum-tube.

(D) Selenium.

(E) High-frequency.

Fig. 11-1. Diodes.

Vacuum-Tube Diodes

The exterior of a vacuum-tube diode is usually a glass tube (sometimes metal) fitted into an insulated base. The interior is a vacuum (to permit ease of electron flow). Suspended in the vacuum are two elements, as shown in the schematic symbol (Fig. 11-2), called a **cathode** and a **plate**. A third element, a **heater**, is mounted close to the cathode.

PLATE
CATHODE

HEATER

Fig. 11-2. Diodes schematic symbol.

Each element is connected to a **pin** (two are for the heater) on the base of the tube. The pins fit into a tube **socket** having terminals to which connections can be readily made.

The Cathode—The cathode is normally a smaller metal cylinder sitting upright in the tube. It is coated with a certain type of material which emits electrons when heated. A voltage (the amount depends on the tube design) applied to the filament wires of the heater raises the temperature of the cathode high enough to cause the coating to "boil off" a cloud of electrons. The electron cloud surrounds the cathode as long as the heater gives off heat.

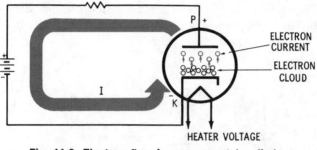

Fig. 11-3. Electron flow in a vacuum-tube diode.

The Plate—The plate (sometimes called an **anode**) is a metal cylinder mounted around the cathode. With a battery connected to the diode as shown in Fig. 11-3, current flows through the tube. Negative voltage on the cathode (symbol K) repels the negative electrons from the cloud toward the plate (symbol P). Positive voltage on the plate attracts the electrons. Current flows from the plate through the external load, through the battery, and back to the tube via the cathode.

If the voltage polarity is reversed, current does not flow. A negative plate repels electrons back toward the cloud, and a positive cathode attracts them in the same direction.

Q11-1. Electrons flow from the _____ to the _____ in a vacuum-tube diode.

Q11-2. When heated, a(an) _____ _____ forms around the cathode.

Solid-State Diodes

In a solid-state diode, current flows through a material rather than a vacuum. There are two general types—one is a **metallic rectifier** and the other a **semiconductor diode**.

Metallic Rectifiers—These are constructed of two different metals tightly pressed together.

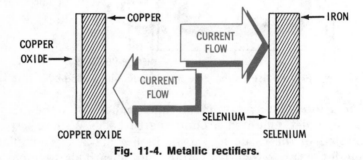

Fig. 11-4. Metallic rectifiers.

A **copper-oxide rectifier** is made of a thick disc or square of copper upon which a thin layer of copper oxide is deposited. As Fig. 11-4 shows, current flows from the copper to the copper oxide if a negative voltage is applied to the copper and a positive voltage to the oxide. If the polarity of the voltage is reversed, a very small amount of current flows in the opposite direction. It is small compared to the amount that flows in the normal direction. In other words, the rectifier has a very low resistance in the direction from the copper to the copper oxide and a very high resistance in the opposite direction.

A **selenium rectifier** has a thick base of iron on which is deposited a thin layer of selenium. Current passes very easily from the selenium to the iron, but a very high resistance path exists in the opposite direction.

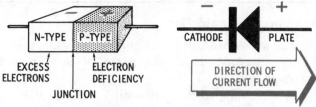

Fig. 11-5. Semiconductor diode and symbol.

Semiconductor Diodes—A semiconductor diode is made of the same materials as those used in transistors, usually **germanium** or **silicon**. The silicon diode is more expensive than the germanium diode, but is able to handle a greater amount of current.

During manufacture, a tiny block, identified as **P-type,** is treated in such a way as to have a deficiency of electrons. Another block is identified as **N-type,** and is treated to have an excess of electrons. Such a diode is often called a **PN junction.**

When joined, a voltage barrier forms at the junction, preventing the electrons in the N material from moving over to the P material. However, when a voltage is applied (as shown in Fig. 11-5) the barrier is overcome and electrons flow from N to P. The schematic symbol used for all solid-state diodes is also shown.

Voltage polarities necessary for current to flow are labeled on the symbol. Current flows through the diode toward the arrowhead.

Q11-3. The arrowhead of a semiconductor symbol corresponds to the _____ of a vacuum-tube diode.

Q11-4. For current to flow in a diode, the cathode must have a _____ voltage with respect to a _____ voltage on the plate.

Q11-5. N-type germanium has a(an) (excess, deficiency) of electrons.

Q11-6. From plate to cathode in a diode is a direction of _____ resistance.

Q11-7. In one type of metallic rectifier, current flows best from _____ to iron.

Q11-8. The letter symbol for a cathode is _____, and for a plate it is _____.

Your Answers Should Be:

A11-3. The arrowhead of a semiconductor schematic symbol corresponds to the **plate** of a vacuum-tube diode.

A11-4. For current to flow in a diode, the cathode must have a **negative** voltage with respect to a **positive** voltage on the plate.

A11-5. N-type germanium has an **excess** of electrons.

A11-6. From plate to cathode in a diode is a direction of **high** resistance.

A11-7. In one type of metallic rectifier, current flows best from **selenium** to iron.

A11-8. The letter symbol for a cathode is **K**, and for a plate it is **P**.

DIODE REACTION TO AC AND DC

It can be easily proved that a diode allows current to flow in one direction and not in the other by constructing the circuit below. A 1.5-volt lamp, an ammeter, and a diode are connected in series across a 1.5-volt cell. (Semiconductor diodes are distinguished from each other by a number-letter designation. The diode recommended for this circuit is a 1N539.)

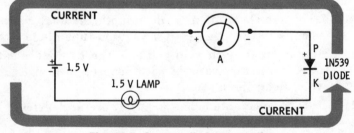

Fig. 11-6. Correct diode connection.

With the connections made as shown in Fig. 11-6, current will flow and the lamp will light. The ammeter will record very close to 200 milliamps. Now reverse the connections of the diode. Will the lamp light? No. And, as indicated in Fig. 11-7, the meter pointer will remain on zero. The plate-to-

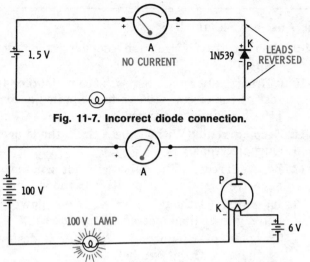

Fig. 11-7. Incorrect diode connection.

Fig. 11-8. Vacuum-tube diodes need two-voltage sources.

cathode resistance of the diode is sufficiently high to prevent a flow of current. A small amount may leak through, but it is not enough to record on the meter.

A vacuum-tube diode is shown in a similar circuit in Fig. 11-8.

Current flows. Note that this particular diode requires a 6-volt source to heat the filament and 100 volts across the cathode and plate. Other vacuum-tube diodes operate with higher or lower voltages, but all require two voltage sources.

Q11-9. In Fig. 11-8 current travels through the lamp from (left, right) to (left, right).

Q11-10. If the vacuum tube and meter have zero resistances and the lamp has a resistance of 150 ohms, how much current will flow?

Q11-11. If the 6-volt battery is disconnected, how much current will flow?

Q11-12. How much current will flow if the 100-volt battery leads are reversed?

Q11-13. With the leads reversed, how much voltage will exist between K and P of the diode?

Q11-14. The plate of a semiconductor diode is designated by the (bar, arrowhead) symbol.

RECTIFYING AC

To **rectify** means to convert ac to dc. A **rectifier** is a device that accomplishes this, and a **rectifying** circuit is one in which it is done. A diode is a **rectifier**.

You know that an alternating voltage increases and decreases in positive voltage during a half cycle and in the next half cycle makes the same changes in negative voltage.

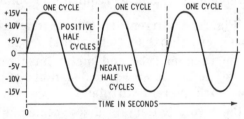

Fig. 11-9. An alternating voltage.

As long as the voltage is generated, the positive and negative half cycles repeat themselves alternately (Fig. 11-9).

Diode Reaction to AC

How does current flow in a diode circuit with ac voltage applied? Remember, current flows in one direction through

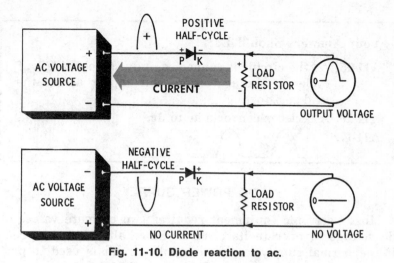

Fig. 11-10. Diode reaction to ac.

a diode only when the plate is more positive than the cathode.

During a positive half cycle, the upper terminal of the source is positive with respect to the lower terminal. This means the lower terminal may be at zero volts, but it is negative when compared to the positive upper terminal. Therefore, the plate of the diode is positive with respect to the cathode. A changing half cycle of current flows.

If a device called an **oscilloscope** is connected across the load resistor, an exact picture of the changing voltage drop will be shown. This changing voltage will look exactly like the waveform at the voltage source.

During the next half cycle, the upper terminal voltage becomes negative with respect to the lower, and the diode plate is negative with respect to the cathode. As you know, current will not flow under these conditions. Since the diode presents a very high resistance in the circuit, all of the source voltage will be dropped across it. No voltage will be displayed across the load resistor.

Q11-15. As the circuit is connected in Fig. 11-10, only the _____ half cycles of the ac voltage will appear across the load resistor.

Q11-16. A rectifier converts _____ to _____.

Q11-17. Draw a diagram showing the waveforms that will appear across R in Fig. 11-10.

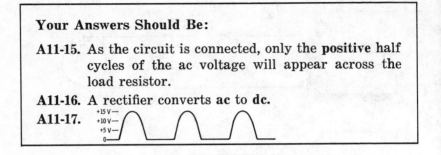
A DC POWER SUPPLY

Most electronic equipment requires two or more values of dc voltage to operate its circuits. Since alternating current is the normal supply, a **power-supply** circuit is used to provide the required dc voltages. Either a solid-state or a vacuum-tube diode is used for the initial conversion.

Filter Circuit

Your diagram shows the output voltage obtained from the diode circuit is dc—current flows in only one direction—but it is not a smooth, nonvarying dc. In fact, it is called a **pulsating dc.** The waveform is a series of pulses.

Filtering Action—The peaks and valleys of the pulsating waveform can be smoothed out by a **filter circuit** (Fig. 11-11). A capacitor can be used to filter (smooth) out some of the changes.

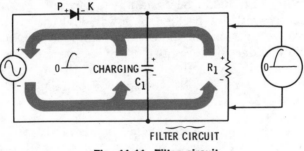

Fig. 11-11. Filter circuit.

As the source voltage rises to maximum positive, current flows through R_1 and the diode. Some of the current also charges the capacitor to the value of the source voltage.

At the instant the applied voltage begins to decrease, C_1 starts discharging, trying to maintain the same voltage level. The discharge path of C_1, however, is through R_1. The current from C_1 cannot flow backward through the diode. Since resistance regulates the time of discharge (5 RC time constants to discharge completely), the discharge time is slow. As shown in Fig. 11-12, the output waveform does not follow the descending curve of the input. It decreases at a much slower rate.

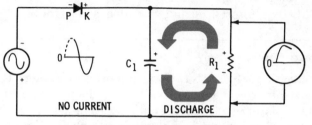

Fig. 11-12. Filter discharging.

During the negative input half cycle, the diode does not allow current to flow, but the capacitor continues to discharge. The discharge current decreases as the capacitor charge grows less. On the next positive swing, the diode does not conduct current until after the input voltage has increased to an amount equal to the charge on C_1 at that instant. The plate must be more positive than the cathode for the diode to conduct. The sequence continues. The resulting output waveshape (in solid lines) is shown in Fig. 11-13. Such an output is called **dc ripple voltage**.

Fig. 11-13. Dc ripple voltage.

Q11-18. The pulse-like waveform developed by a diode circuit is called _____ _____.

Q11-19. The changing voltage pattern made by these pulses can be smoothed out by a _____ circuit.

Q11-20. The filter capacitor begins to _____ as soon as the positive voltage input begins to decrease.

Q11-21. The filtered output is called a _____ _____.

Full-Wave Power Supply

The circuit in Fig. 11-13 is a **half-wave** rectifier. It allows only half the ac wave (positive half cycles) to appear across the load resistor. **Full-wave** (both positive and negative half cycles) rectification can be obtained with the switching action of the diodes in the circuit in Fig. 11-14.

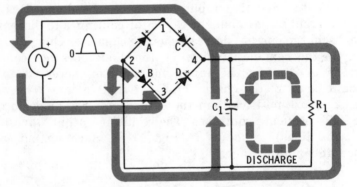

Fig. 11-14. Full-wave rectifier—positive half cycle.

On the positive half cycle, current leaves the lower terminal of the ac voltage source and enters terminal 3 of the 4-diode network (called a **bridge**). The bridge is positive to negative from top to bottom because of the source polarity. Diode B has a negative cathode and a positive plate, but diode D has reverse polarity across it. Current must, therefore, flow through diode B to terminal 2. This current charges C_1 and flows through R_1 to terminal 4. Because of polarities, this current must flow through diode C. Diodes D and A are of the wrong polarity.

C_1 charges as the voltage increases and discharges during the voltage decrease, just as in the half-wave circuit.

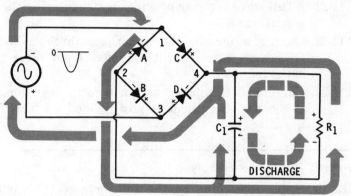

Fig. 11-15. Full-wave rectifier—negative half cycle.

During the negative half cycle, the polarities of the voltages on the four diodes are reversed. Current leaving the upper end of the source arrives at terminal 1. The voltage on diode C is of the wrong polarity, but diode A will conduct. Current leaves terminal 2, and then follows the same path as the positive half-cycle current. Capacitor C_1 has just begun to discharge; the rising current restores the charge to full voltage. The remainder of the current flows through R_1 in the same direction as the positive half-cycle current did. At terminal 4, only diode D has the correct voltage polarity to conduct. Current flows through diode D to terminal 3 and the ac source.

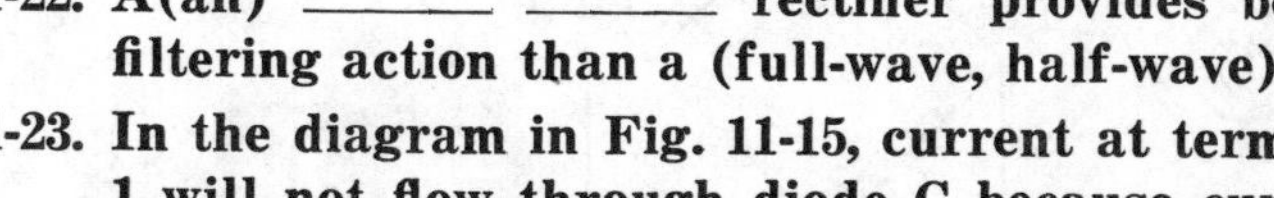

POS. NEG. POS. NEG. POS. NEG.

Fig. 11-16. Full-wave ripple voltage.

Q11-22. A(an) _____ _____ rectifier provides better filtering action than a (full-wave, half-wave).

Q11-23. In the diagram in Fig. 11-15, current at terminal 1 will not flow through diode C because current will not pass from _____ to _____.

Q11-24. Current at terminal 4 will not go through diode C because its cathode is _____ and its plate is _____.

Improving the Power Supply

The dc ripple remaining on the full-wave rectifier output may be satisfactory dc voltage for some equipment but not for others. The output can be made still smoother by improving the filtering action (Fig. 11-17).

By adding another capacitor in parallel with the load resistor and another current-limiting resistor in series with the discharge path, the filter network can reduce more of the ripple.

Both capacitors are charged and recharged by the positive and negative currents switched into the filter by the bridge. Both capacitors discharge together through R_1 as C_1 did previously. Resistor R_2 aids by limiting the flow of current through the filter.

Further improvement to the filtering action can be made by replacing R_2 with an **iron-core coil.** Such a coil is wound on a bar of iron. The reaction of a coil (inductor) to ac is, as you recall, one of opposing changes in current. Magnetic

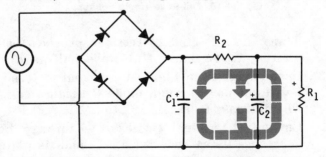

Fig. 11-17. A capacitor-resistor-capacitor filter.

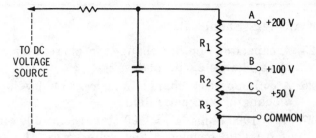

Fig. 11-18. The load resistor can supply different voltages.

fields, reinforced by the iron core, smooth out the ripple by preventing the changes from occurring.

Load Resistors

Some electronic equipment requires two or more values of dc voltage for proper operation. These voltages can be selected from the load resistor.

Suppose that the dc requirements of the rest of the equipment were +200, +100, and +50 volts. A power supply can be selected or designed to produce a current large enough to cause a drop of at least 175 volts across R_1.

Either three series resistors of the correct values, or a bleeder resistor capable of being tapped at the desired values, can be used. The drawing in Fig. 11-18 shows a bleeder resistor symbol.

By making connections to terminals A and Common, 200 volts will be available (for the plate of a vacuum tube, for example). To obtain 100 volts, the bleeder is tapped at the halfway point to obtain half the total voltage. For 50 volts, the resistance is tapped halfway between the 100-volt point and the common terminal.

Q11-25. A capacitor opposes changes in current by storing a _____ on its plates.

Q11-26. A coil opposes a change in current by developing a changing _____ _____.

Q11-27. Assume in Fig. 11-18 that the bleeder must be replaced with three separate resistors. If you know that 0.1 amp flows through the bleeder to produce a total of 200 volts, what is the value of R_1, R_2, and R_3?

WHAT YOU HAVE LEARNED

1. Diodes are constructed in the form of vacuum-tube or solid-state devices.
2. All vacuum-tube diodes have a cathode and plate. The plate must be positive with respect to the cathode to permit current flow. Current will not flow in the reverse direction (plate to cathode) regardless of the polarity.
3. The plate and cathode in a vacuum-tube diode are housed in a vacuum. Heating the cathode causes it to emit electrons.
4. Solid-state diodes (metallic rectifiers and semiconducttors) allow current to flow in one direction under conditions of proper polarity because of the materials from which they are made.
5. A diode converts ac to dc because it forms a one-way street for current. A device that does this is called a rectifier.
6. Output from a rectifier is pulsating dc. To smooth out the pulsations, a full-wave rectifier can be used. By adding a filter circuit—capacitors and resistors, or capacitors and a coil—the ripple can be made relatively smooth.
7. The smooth dc voltage can be taken from the power supply in desired values by tapping a bleeder resistor.

12

How Vacuum
Tubes Work

what you
will learn

You are now going to
learn the basic principles
about a device you have
seen before—a vacuum
tube. One member of the
vacuum-tube family is a diode. The vacuum-tube group
has several other members, many of which amplify and
reshape the signals passing through your radio and television receivers. You will become familiar with the other
types of tubes, learn how they work, and become acquainted with how some of them are used in circuits.

WHAT ARE VACUUM TUBES?

You should know at least part of the answer to this question. If a diode has two active elements (cathode and plate)
suspended in an evacuated (vacuum) enclosure (tube), then
all other vacuum tubes probably have the same number of
elements or more.

That is exactly the situation. A **triode** is a vacuum tube
with **three** active elements. A **tetrode** has **four**, and a **pentode**
has **five**. There are other tubes with more elements, and still
others with special elements. However, they all obey the same
principles that will be discussed here.

Each vacuum tube has a filament to heat the cathode. Although the filament itself is not considered an active element,
it performs a necessary function—it produces the heat to boil
a cloud of electrons away from the cathode.

HOW DOES A VACUUM TUBE WORK?

As you learned in the study of a diode, the cathode emits electrons and the plate receives electrons. Have you wondered how a diode is constructed to perform these functions?

Vacuum-Tube Fundamentals

Although the heaters, cathodes, and plates of the various kinds of vacuum tubes may be built to slightly different dimensions and shapes, the operating principles of all vacuum tubes are the same. The tube elements are located in a vacuum to eliminate any air molecules that would retard the free flow of electrons to the plate.

Heater and Cathode—The heater (filament) is a fine wire which, when energized, raises the temperature of the cathode to the desired emitting level. Heater-cathode combinations are of two types.

Part A in Fig. 12-1 shows a filament-cathode. The single wire serves as both the filament and cathode; it is made of a material (normally coated or uncoated tungsten) that emits electrons when hot.

Part B shows a heater-cathode combination in which there are two separate elements. The cathode, coated with an electron-active substance, fits like a sleeve over the filament.

The filament-cathode requires less current to heat it to an electron-emitting temperature. This type of cathode works best when connected to a dc voltage source. If energized by ac, tube current would vary with the ac alternations. But because ac current is more easily supplied, most of the tubes used in electronic equipment are of the heater-cathode type. The

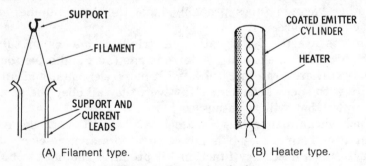

(A) Filament type.   (B) Heater type.

Fig. 12-1. Types of cathodes.

heater and the cathode are electrically insulated from each other so that the alternating current flowing through the heater does not affect the tube current.

Plate—The plate is usually a metal cylinder surrounding the cathode. Construction details and the accepted diode tube symbol are shown in Fig. 12-2.

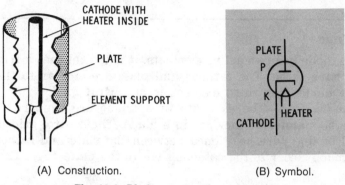

(A) Construction. (B) Symbol.

Fig. 12-2. Diode construction and symbol.

Cathode Temperature—The number of electrons that form in a cloud around the cathode depends on the cathode temperature. The temperature of the cathode is controlled by the heat generated by the filament. There is an upper limit, however, beyond which a further increase in temperature will not cause an increase in electron emission. This is called **cathode-temperature saturation.**

Plate Voltage—The plate attracts electrons from the electron cloud when the plate voltage is positive with respect to the cathode. By raising and lowering the plate voltage, a greater or fewer number of electrons will be drawn from the cathode. This increases or decreases the value of plate current (tube current). The upper limitation, where a further increase in plate voltage will not attract any additional electrons, is called **plate-current saturation.**

Q12-1. A cathode emits _____.

Q12-2. Plate current flows when the plate is _____.

Q12-3. Plate current can be increased by increasing the _____ on the _____.

The Triode

If a diode has two active elements, it is logical that a triode must have three. The cathode and plate are similar to those in the diode. The third element is a **control grid.**

Construction—The elements in a triode are supported in the same manner as they are in a diode. The control grid is a spiral of fine wire positioned between the plate and cathode. It is much closer to the cathode than to the plate (Fig. 12-3).

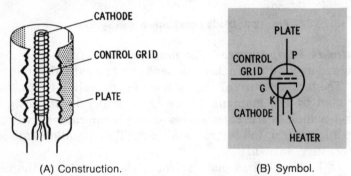

(A) Construction.　　　　　(B) Symbol.

Fig. 12-3. Triode construction and symbol.

Operation—Plate current in a diode is determined by cathode temperature and plate voltage. In a triode, the voltage of the grid with respect to the cathode also controls the amount of plate current.

Because of its nearness to the cathode, the grid has greater control over the number of electrons reaching the plate than the plate does itself. Large changes in plate voltage cause only small changes in plate current. But small changes in grid voltage cause large changes in cathode-to-plate current.

In fact, if the grid is made sufficiently negative with respect to the cathode, the flow of current will be stopped or cut off.

232

The lowest voltage at which this occurs is called the **cutoff bias.** Bias voltage is the normal voltage difference between cathode and control grid. The grid is usually more negative than the cathode.

A Triode Circuit—In Fig. 12-4 is a circuit which shows how a triode functions.

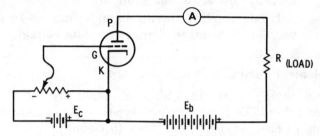

Fig. 12-4. A triode amplifier.

Battery E_B is connected in such a way that the plate is positive with respect to the cathode. In an actual circuit the positive voltage for the plate is normally obtained from a power supply. Battery E_C places a negative voltage on the grid with respect to the cathode. A resistor takes the place of this battery in an actual circuit.

When the grid voltage is sufficiently negative to cut off the plate current, no current will flow through the load resistance. If made less negative, the grid will allow some current to flow and a voltage will be developed across R (load). If made even less negative, more plate current will flow and a greater voltage will be dropped across the resistor.

Assume that the change in grid voltage in the last two steps is 2 volts (from −6 to −4). Also assume that the change in the plate-resistor voltage is a total of 60 volts. This means that the grid-voltage change has been amplified 30 times (60/2). An ac signal on the grid would cause the same amount of amplification. This is how a triode amplifies.

Q12-4. The control grid is mounted closer to the (cathode, plate) than to the _____.

Q12-5. The voltage on the control grid that stops plate current is called _____ bias.

Q12-6. A triode amplifies because _____ changes in grid voltage cause _____ changes in plate voltage.

233

Multielement Tubes

Tubes having more than three elements are called **multi-element** tubes. The most common types in this group are the tetrodes (4 elements) and pentodes (5 elements).

Tetrodes—Triodes are very good amplifiers of low-frequency signals. At high frequencies, however, such as those used in radio and television, a triode distorts (changes the form of) a signal during amplification. The distortion is caused by the capacitance that exists between the plate and grid. Since these two elements are conductors separated by an insulator (vacuum), a capacitor is formed. Capacitance also exists between the other elements, but has less effect on the signal.

Part A in Fig. 12-5 shows the **interelectrode** (between elements) **capacitance** in a triode. Plate voltages are fed back to the control grid through this route, causing distortion. In a tetrode, the extra grid (called a **screen**) between the control grid and plate reduces the interelectrode capacitance and can be used to divert the feedback voltage to **ground** through a capacitor. Note the tetrode symbol in Part B. **Ground** is a wire (or the chassis) which serves as a common conductor for, or connection to, other components.

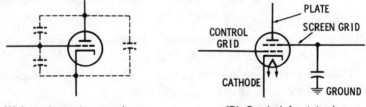

(A) Interelectrode capacitance.   (B) Symbol for tetrode.

Fig. 12-5. A tetrode is better for high frequencies.

Pentodes—The fifth element in a pentode alleviates another problem encountered in vacuum tubes—the problem of **secondary emission.** Many of the electrons making up the plate current strike the plate with sufficient velocity to release other electrons from the plate material and bounce them back into the space between the screen grid and plate.

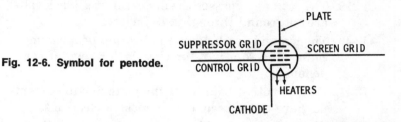

Fig. 12-6. Symbol for pentode.

The fifth element of the pentode is called a **suppressor grid.** This element is usually connected internally to, and has the same potential as, the cathode. In other words, the suppressor grid is negative with respect to the plate. Spacing between the turns of wire of the suppressor grid is wide enough for plate current to pass through, but yet sufficiently close enough to repel the negative secondary electrons back to the plate.

Q12-7. _____ _____ between the plate and grid of a triode causes signal distortion.

Q12-8. The fourth element of a tetrode is called a _____ _____.

Q12-9. The screen grid bypasses the distorting feedback voltage to _____ through a capacitor.

Q12-10. The _____ symbol is an indication of a common connection to a wire or chassis for several components.

Q12-11. Electrons that bounce off the plate due to current flow are called _____ _____ electrons.

Q12-12. The _____ grid of a pentode repels electrons back to the plate.

Q12-13. The _____ grid is tied to the cathode.

VACUUM-TUBE CIRCUITS

There are thousands of different vacuum-tube circuits. No one could hope to learn how each circuit works by memorizing their operating details. However, you can analyze how they work by applying the principles of electricity and electronics. You have already acquired most of the fundamental principles, but not at the depth required to be an expert. See if you can apply some of these principles to the circuits that follow.

Full-Wave Vacuum-Tube Power Supply

The circuit in Fig. 12-7 performs the same rectification function as did the diode full-wave power supply in the last chapter.

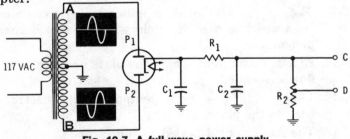

Fig. 12-7. A full-wave power supply.

The transformer steps 117 volts ac up to the value desired on the secondary. Each terminal of the secondary is connected to a separate plate of the **twin diode**. The secondary winding of the transformer is center-tapped to ground.

The filter network is formed by C_1, R_1, and C_2. By storing a charge during each ac half cycle, and discharging slowly through the resistors, the capacitors smooth out the ripple in the pulsating dc. Bleeder resistor R_2 provides selected dc voltages for other vacuum tubes.

Note that C_1, C_2, and R_2 each have one side connected to ground. The ground symbol indicates a common connection for all components terminated in this manner.

The principle of a center-tapped transformer winding is the manner in which ac voltages appear on either end of the winding. The first half cycle appearing at point A is positive with respect to ground. This makes the center tap more positive than point B, or point B is, in effect, swinging in a negative direction.

When point A is positive, point B is negative. Plate P_2 is cut off, but P_1 conducts. Current travels through the upper half of the secondary to ground, up through R_2 (charging the capacitors on the way), through R_1, and then to the cathode. The top of R_2 is positive with respect to ground.

When point A swings negative, point B swings positive.

Q12-14. When a negative half cycle is developed from point A to ground (P_1, P_2) conducts.

Q12-15. With current from cathode to P_2, current flows through R_2 from (top to bottom, bottom to top).

Q12-16. When current leaves P_1, it will enter R_2 at the (top, bottom) and leave at the (top, bottom).

Q12-17. When P_2 conducts, C_1 will charge by storing excess electrons on the (top, bottom) plate.

Q12-18. The vacuum tube in the circuit is a _____ _____.

Q12-19. The voltage at point A would be measured from point A to _____.

Q12-20. If the voltage output at point C is 200 volts, what will be the voltage at point D if this point is ⅗ of the resistance above ground?

Triode Amplifier

The circuit in Fig. 12-8 is that of a triode amplifier. The B+ voltage of 200 volts is obtained from the load resistor (bleeder) of a power supply. The control grid has a fixed bias of −2 volts. Since there is a common connection through ground, the grid is two volts negative with respect to the cathode.

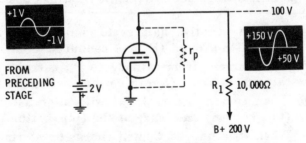

Fig. 12-8. A triode amplifier.

With this bias, plate current is 0.01 amp (10 mA). The cathode-to-B+ path contains R_1 and r_p as resistances in series. R_1 is the load across which the voltage for the next circuit is developed; r_p is a variable resistance existing between cathode and plate. This variable resistance is known as the **plate resistance.** When plate current increases, r_p decreases; when plate current decreases, r_p increases.

With 0.01 amp flowing, there will be a 100-volt drop across R_1 ($E = IR$). Since B+ is 200 volts, there will be a 100-volt drop across r_p ($200 - 100$). This is the same as saying that the voltage from plate to cathode is 100 volts.

The ac signal on the grid swings one volt positive. Added to the -2-volt bias voltage, the grid is now at -1 volt with respect to the cathode. Plate current increases to 0.015 amp. The drop across R_1 is now 150 volts, leaving 50 volts across r_p. A change of 1 volt on the grid has changed the plate voltage from 100 volts to 50 volts.

When the ac signal swings one volt negative, grid bias becomes -3 volts. Plate current decreases to 0.005 amp. The voltage drop across R_1 is now 50 and across r_p, 150. Once again a change of 1 volt on the grid has caused a change of 50 volts on the plate. This means that the gain of the tube is 50.

Q12-21. When plate current increases, r_p _____.

Q12-22. Gain of an amplifier is determined by dividing the change in _____ _____ by the change in

_____ _____.

WHAT YOU HAVE LEARNED

1. A diode has two active elements—a cathode and a plate. When heated, the cathode emits electrons. The plate draws current when it is more positive than the cathode.

2. A triode has a cathode, plate, and control grid. The control grid regulates the amount of plate current. A small change in grid voltage causes a large change in plate current and voltage.

3. The additional grid in a tetrode is called the screen. It eliminates the distortion effects of interelectrode capacitance.

4. A pentode has a suppressor grid that returns secondary electrons to the plate.

5. The ground symbol indicates a common wire or chassis for all components terminated to ground.

6. A full-wave vacuum-tube power supply uses a twin diode and a transformer center-tapped-to-ground secondary as input voltage.

7. Increasing plate voltage increases plate current until the limit of plate-current saturation is reached.

8. Varying cathode temperature varies the supply of available electrons and, therefore, plate current. Maximum electrons will be available at cathode-temperature saturation.

TO THE READER

Thus far this volume has introduced you to many of the basic concepts of electricity and electronics. In fact, you should now be familiar with all of the basic principles upon which this science is based. Although you will need to learn more about these principles before you can become an accomplished technician, what you know now will make such study understandable and meaningful.

To ensure that your grasp of these principles is sound, the remainder of this volume will be devoted to the application of these concepts to actual circuits in typical equipment. It will be more than just a review. The many components that you may have studied will be tied together in describing how actual equipment works. In addition, you will become acquainted with the manner in which your radio and television receivers function.

13

Basic Circuit Actions

what you will learn

It is important that you learn what an electrical or electronic circuit is. You will now learn to recognize the basic elements every circuit must have. When you complete this chapter you will be able to examine a circuit and determine how it works. In addition, you will learn the simple fundamentals used to determine how any circuit works.

INTRODUCTION

In the preceding chapters you have learned a great deal about electricity and electronics. If you have performed the experiments, you have learned even more.

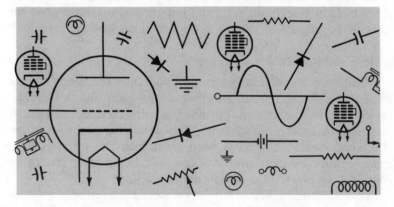

Fig. 13-1. Circuits contain many parts.

The remaining chapters of this volume build upon what you have learned by showing you what electronic circuits are and how they are used in electronic equipment.

ELECTRICITY AND ELECTRONICS

You will read or hear many definitions for electricity and electronics which will seem to establish a difference between the two terms. Are they really different? The truth is, electricity and electronics are far more similar than they are different.

Electrical Circuits

In an electrical circuit, current from a voltage source flows through conductors to an electrical device. In passing through the device, current causes it to operate—a lamp lights, a motor rotates, a doorbell rings, an oven heats, etc. The electrical device, whatever it may be, must be part of a circuit connected to a voltage source. Compare this to an electronic circuit.

Electronic Circuits

In an electronic circuit, current from a voltage source flows through conductors and electronic components to perform a desired electronic function. For example, radio and television receivers contain many electronic circuits. In a radio, the functions of each circuit are such that the set reproduces a sound transmitted by a broadcast station many miles away. Circuits in a television set function in a similar manner and make it possible for you to see as well as hear a broadcast.

Basic Fundamentals

A radio or television set is plugged into the same voltage source as a lamp, motor, refrigerator, or any other electrical device. Current and voltage make no distinction between an electrical or electronic circuit. They react the same in either. The components that have been built into the circuit(s) determine how current and voltage will be used to make the electrical or electronic device operate.

The diagram in Fig. 13-2 expresses this concept by showing the relationship of the three elements contained in any circuit. If you accept the concept this illustration reveals and always remember it, you will have no difficulty in learning the electrical or electronic theory required to become a good technician.

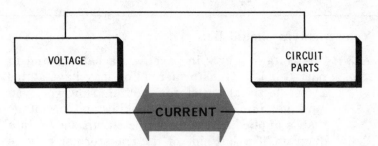

Fig. 13-2. The electrical elements in any circuit.

What does the illustration say? It states that any circuit contains only three factors—voltage, current, and circuit parts —which influence its operation.

Voltage, as you know, is an electrical pressure that causes current to flow under proper conditions. Current flows if there is a closed loop (complete path) from one side of the voltage source through the circuit components to the other side. The amount and type of voltage to be applied and how much current will flow is dependent on the type and value of components used in the circuit. As an example, you have seen circuits similar to those shown in the schematic diagrams in Fig. 13-3.

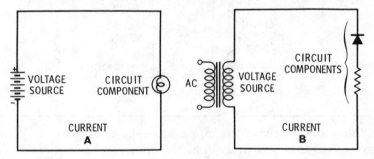

Fig. 13-3. Circuit examples.

Q13-1. What is the voltage source in part A in Fig. 13-3? In part B?

Q13-2. What is the circuit component in part A? What are they in part B?

Q13-3. Which way will current flow (clockwise or counterclockwise) in parts A and B?

Q13-4. Which part (A or B) contains a circuit that is a closed loop?

Your Answers Should Be:

A13-1. The voltage source in part A is a **battery** and in part B it is a **transformer**. (You may have stated ac or electrical outlet for part B. However, you must get into the habit of looking at the voltage that is applied across a specific circuit shown in a diagram. The left side of the transformer may be plugged into a 117-volt ac outlet, but it is in another circuit consisting of the primary winding and the outlet. The circuit shown contains the voltage source and the circuit component. Remember to look for the specific voltage that is applied to an individual circuit.)

A13-2. The component in part A is a **lamp.** In part B the components are a **diode** (upper) and **resistor** (lower).

A13-3. Current will flow **counterclockwise in part A** and **clockwise in part B.**

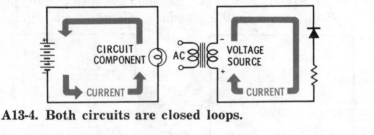

A13-4. Both circuits are closed loops.

You were not expected to give as long an answer as those shown above. If you arrived at the specific answers correctly, you have shown that you can recall and apply the information studied in other chapters of this volume.

The questions and explanations were included to underline a significant point—whatever happens in any circuit depends on the effect circuit components have on its voltage and current. This may sound like a very simple, easily understood statement, but those who do not study circuits with this simplicity in mind will find them difficult. Those who approach every circuit and resolve its complexities in terms of this simple, always reliable statement will have no trouble whatsoever.

ANALYZING ELECTRONIC CIRCUITS

Using the approach stated in the preceding paragraph, see if you can follow the analysis of a basic amplifier circuit (Fig. 13-4). The circuit is similar to one of those used in preceding chapters.

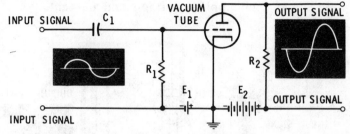

Fig. 13-4. A vacuum-tube amplifier.

Circuit Function

This circuit uses a vacuum tube. The function of the circuit is to amplify (increase) the voltage of the input signal, as shown in the difference between the input and output waveforms. Disregarding the input and output signals for a moment, you can find two voltage sources in this circuit—there are two battery symbols. The actual circuit will probably not have batteries; the symbols merely show that there is a dc voltage source across the points indicated.

Q13-5. The three basic factors of any circuit are _____, _____, and _____.

Q13-6. The best method to use in analyzing any circuit is to determine the effect that circuit components have upon applied _____ and _____.

Q13-7. Redraw the vacuum-tube amplifier circuit and show the path and direction of current through the closed loop that includes E_2.

Q13-8. The three active elements in the vacuum tube above are _____, _____, and _____.

Q13-9. The purpose of the circuit is to _____ the input signal.

Q13-10. The two battery symbols are used to indicate _____ _____ is being applied.

A13-5. The three basic elements of any circuit are **voltage, current,** and **components.**

A13-6. The best method to use in analyzing any circuit is to determine the effect that circuit components have upon applied **voltage** and **current.**

A13-7.

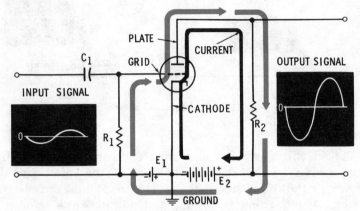

You had two choices for the current path. They are represented as solid and broken lines above. Although the solid line is correct, do not feel badly if you chose the other.

A13-8. The three active elements in the vacuum tube are **control grid, cathode,** and **plate.**

A13-9. The purpose of the circuit is to **amplify** the input signal.

A13-10. The two battery symbols are used to indicate **dc voltage** is being applied.

Control by the Grid

Current (the solid line) flows through the vacuum tube (a triode) if its plate is positive with respect to its cathode. This is the purpose of E_2. The amount of current that flows can be controlled by the voltage on the control grid with respect to the cathode. In fact, a small change in grid-to-cathode voltage causes a large change in plate current.

The grid-to-cathode voltage is negative and, if sufficiently high, will stop current flow altogether. As this voltage is made less negative more and more current will flow.

The input to the triode amplifier circuit has the same shape as an ac-voltage waveform. It appears on the grid as a voltage also. Capacitor C_1 and resistor R_1 play a part in placing this signal voltage on the grid. The line through the center of the input waveform is called a **reference line**. (Voltage must always be thought of as being with reference to, or with respect to, some other point in the circuit.) In this case the reference line refers to the dc grid voltage (bias). The part of the waveform that is above the line is positive voltage, and the part below the line is negative. Since it is ac, the voltage of the signal is regularly changing from positive to negative.

The purpose of E_1 is to establish a uniform negative voltage between the grid and cathode. This makes the grid negative with respect to the cathode. In this circuit, current will not flow from the grid to the cathode. The changing voltage of the ac waveform is also on the grid, subtracting from or adding to the voltage of E_1. When the signal voltage is going positive, it subtracts from the voltage of E_1. For example, if E_1 was −1.5 volts (negative from grid to cathode) and if the signal was +0.5 volt at a given instant, voltage on the grid would be reduced to −1 volt (the grid still negative with respect to the cathode).

Q13-11. The control grid is (negative, positive) with respect to the cathode.

Q13-12. What will be the voltage on the grid when the signal voltage is −0.5 volt (E_1 is −1.5 volts)?

Q13-13. E_2 makes the plate _____ with respect to the cathode.

Q13-14. The amount of plate current that flows is controlled by the _____ on the _____ _____.

Q13-15. In a triode amplifier, a small change in grid voltage causes a (small, large) change in plate current.

Q13-16. The purpose of E_1 is to make the _____ negative with respect to the _____.

Change in Plate Voltage

As shown in Fig. 13-5, the voltage on the grid changes in accordance with the changing voltage of the signal. Current through the tube changes in a like manner—it increases when the signal rises in the positive direction. This is because the negative repelling voltage of the grid is being decreased. When the signal increases in the negative direction, it adds to the negative voltage on the grid, causing plate current to decrease.

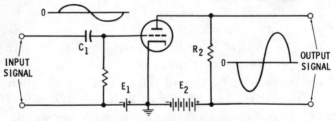

Fig. 13-5. Amplifying a signal.

The changing current of the tube passes through R_2 on its return to voltage source E_2, causing the voltage across R_2 to change in the same manner as the changes of the signal voltage. Since a small change in grid voltage causes a large change in tube current, the changes in output voltage across R_2 are greater than the corresponding input changes on the grid. Thus, the signal has been **amplified.**

Circuit Analysis Summary

The entire explanation or understanding of this circuit is based on the effect the circuit components have on current and/or voltage. This is true of any circuit. You should have had very little difficulty in following the explanation even with your limited knowledge of electricity. The reason, of course, is that everything was explained in terms of changes in voltage or current with respect to the components of the circuit.

You may have been able to follow the explanation but you still may not fully understand exactly how the circuit works. To achieve this understanding and to become a good technician requires a knowledge of how the circuit components cause current and voltage changes to take place.

Many pages are devoted to this explanation in the remaining volumes in this set or in similar texts on electronics. The effect that vacuum tubes, resistors, capacitors, and even voltage sources have on a signal moving through a circuit requires a great deal of careful explanation. If you will remember to always relate the detailed descriptions to the effect they have on voltage and current changes, you will have no trouble.

Q13-17. In a triode amplifier, the (negative, positive) terminal of a dc voltage source is applied to the control grid.

Q13-18. The (input, output) signal causes grid voltage to vary.

Q13-19. Grid voltage regulates the amount of plate current by (attracting, repelling) electrons in the tube.

Q13-20. The circuit on the opposite page is called a(an) _____ because it increases the voltage of the input signal.

Q13-21. The plate of the tube is kept at a (higher, lower) (negative, positive) voltage than the cathode.

Q13-22. The circuit amplifies the input signal because (small, large) changes in grid voltage cause (small, large) changes in plate current.

Q13-23. The changes in plate current cause (small, large) changes in plate voltage.

Your Answers Should Be:

A13-17. In a triode amplifier, the **negative** terminal of a dc voltage source is applied to the control grid.

A13-18. The **input** signal causes grid voltage to vary.

A13-19. Grid voltage regulates the amount of plate current by **repelling** electrons in the tube.

A13-20. The circuit is called an **amplifier** because it increases the voltage of the input signal.

A13-21. The plate of the tube is kept at a **higher positive** voltage than the cathode.

A13-22. The circuit amplifies the input signal because **small** changes in grid voltage cause **large** changes in plate current.

A13-23. The changes in plate current cause **large** changes in plate voltage.

CIRCUIT COMPONENTS

How many different circuit components are there? If this question is worrying you, you are worrying needlessly. There are only three major components, or parts (Fig. 13-6).

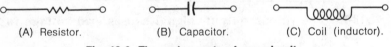

(A) Resistor.　　　　(B) Capacitor.　　　　(C) Coil (inductor).

Fig. 13-6. The major parts of any circuit.

Resistors, Capacitors, and Coils

All circuits, regardless of their complexity, contain at the most only three different kinds of parts—**resistors, capacitors,** and **coils** (often called **inductors**).

The effect that a resistor has on current or voltage is measured in terms of its **resistance,** a term with which you are already familiar. The effect of a capacitor is measured in **capacitance.** The effect of a coil is called **inductance.** The effect each has on voltage or current depends on whether it is dc or ac, and, if ac, how rapidly the voltage or current is changing. But each effect—resistance, capacitance, or inductance—is based on a few easily learned principles.

Circuit Applications

The illustration in Fig. 13-7 shows that an input signal is converted to that shown at the output because of the effect of circuit resistance, capacitance, and inductance on the signal as it passes through the circuit. By this manner, the operation of any circuit can be explained. The **ground** symbol shown in the illustration is normally used as a reference point for zero voltage.

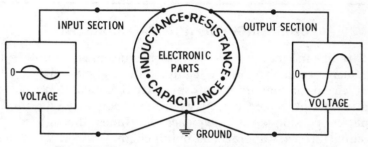

Fig. 13-7. The elements of any electronic circuit.

You might think that the symbol for a vacuum tube or transistor does not look like an inductance, capacitance, or resistance. You are correct. However, the way a vacuum tube or a transistor operates can be explained by how it reacts in terms of resistance, inductance, or capacitance when current is passing through it or when voltage is applied to it.

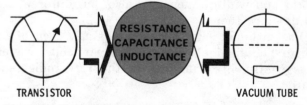

Fig. 13-8. Three different electrical factors in a circuit.

Q13-24. The three different electrical factors in a circuit are _____, _____, and circuit components.

Q13-25. The three different types of circuit components are _____, _____, and coils.

Q13-26. Operation of a vacuum tube can be explained in terms of _____, _____, and _____.

CHANGING VOLTAGE AND CURRENT

Circuits in electronic equipment are designed to obtain the performance desired of the equipment. A signal entering the first circuit is converted into an output signal that becomes the input to the next circuit where it is converted again. The input-conversion-output sequence continues through all the circuits until a waveform is obtained that will cause proper operation of the output device.

Voltage and Current Waveforms

Since the exchange between circuits is accomplished by voltage and/or current, a means of describing a waveform (signal) becomes very important. Like any other object, a waveform has dimensions. A sheet of paper, for example, is so many inches wide by so many inches long. A waveform has height and width dimensions also, but different units are used to describe them.

The illustration in Fig. 13-9 shows a single cycle of a voltage waveform. The ac **sine wave,** as this particular waveform is called, is continually changing at the rate indicated by its curvature.

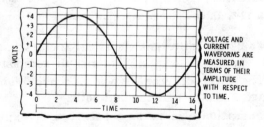

VOLTAGE AND CURRENT WAVEFORMS ARE MEASURED IN TERMS OF THEIR AMPLITUDE WITH RESPECT TO TIME.

Fig. 13-9. A voltage sine wave.

Circuit Applications

Normal presentation in equipment diagrams indicates the amplitude of a waveform in terms of its maximum values.

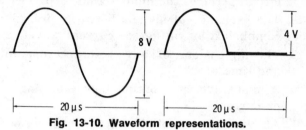

Fig. 13-10. Waveform representations.

Part A in Fig. 13-10 shows the same sine wave as in Fig. 13-9. The dimensions are 8 volts from its positive peak to its negative peak. The time duration of one cycle is 20 μs (microseconds). A microsecond is one-millionth of a second. Part B shows the same waveform after it has been rectified (using a diode circuit, for example). The single peak remaining is 4 volts from zero to maximum positive. The time duration of the cycle is still the same 20 μs.

Understanding the dimensions of a waveform is very important. Waveform representations of signals are used constantly in electronics, since a vast amount of information about a signal can be put into this picture form. Amplitude and time values allow you to describe specifically what the voltage or current will do in a circuit.

Q13-27. In electronic equipment, the output signal of one circuit becomes the _____ _____ for the next circuit.

Q13-28. The exchange of signals between circuits is accomplished by _____ or _____ waveforms.

Q13-29. The dimensions of a waveform are _____ and _____.

Q13-30. A sine wave is a continuously (steady, changing) voltage or current.

Q13-31. In the diagram on the opposite page, what is the amplitude of voltage at time increment 4?

Q13-32. What is its value at time increment 8?

Q13-33. What is the amplitude at time increment 12?

Your Answers Should Be:

A13-27. In electronic equipment, the output signal of one circuit becomes the **input signal** for the next.

A13-28. The exchange of signals between circuits is accomplished by **voltage** or **current** waveforms.

A13-29. The dimensions of a waveform are **amplitude** and **time**.

A13-30. A sine wave is a continuously **changing** voltage or current.

A13-31. At time increment 4, voltage has risen to **+4 volts**.

A13-32. At time 8, it is **zero volts**.

A13-33. At time 12, voltage has decreased to **−4 volts**.

Amplitude and Frequency

As you have learned, waveforms can be described by their time and amplitude dimensions. How is this done?

Time Dimension—Time is the horizontal dimension of a waveform. It is usually represented in terms of seconds, milliseconds (1/1000 of a second), or microseconds (1/1,000,000 of a second).

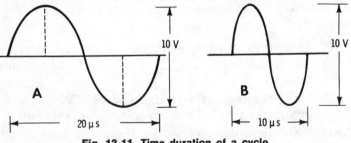

Fig. 13-11. Time duration of a cycle.

The time line for a waveform usually represents the duration of one cycle. In part A in Fig. 13-11, it is 20 μs and in part B it is 10 μs. From this, the duration of a portion of a cycle can be determined. In part A, a half wave (half of a full cycle) is 10 μs. A quarter wave (fourth of a full cycle) in part A is 5 μs.

Frequency—Since a waveform cycle repeats itself continuously, its frequency can be determined. The frequency of a signal is the number of times that it repeats itself in a certain period of time, usually one second. If the time duration for one cycle is one second, the signal repeats itself once each second. Its frequency, then, would be one hertz. If one cycle is 1/10 of a second in duration, it repeats itself 10 times in one second, resulting in a frequency of 10 hertz (cycles per second). As you have already determined, the arithmetic expression to find frequency is

$$\text{Frequency} = \frac{\text{one second}}{\text{time duration of one cycle}}$$

If the time duration is expressed in milliseconds or microseconds, the top and bottom values of the right side of the expression must be expressed in the same units of time. In other words, both top and bottom values must be either in seconds, milliseconds, or microseconds. Failure to have these values in the same units of time is a common source of error when solving this type of problem (Fig. 13-12).

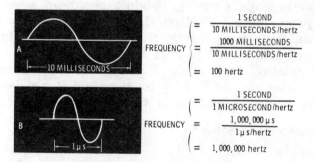

Fig. 13-12. Determining frequency.

Q13-34. The time dimension of a waveform is measured from (left to right, bottom to top).

Q13-35. There are _____ microseconds in a second.

Q13-36. There are _____ microseconds in a millisecond.

Q13-37. If a full cycle is 60 milliseconds, what is the duration of a quarter cycle?

Q13-38. What is the frequency of a 100-millisecond cycle?

Q13-39. What is the frequency of a 0.001-second cycle?

WAVEFORM APPLICATIONS

The time and amplitude characteristics of a waveform allow it to be described in precise terms.

Oscilloscope

An **oscilloscope** is used to obtain a picture of a waveform at test points in a circuit. Even though the signal is constantly changing, controls on this instrument permit the waveform to be presented almost as if it were drawn on paper. From the presentation, the waveform can be evaluated in terms of its amplitude and time characteristics and it can be determined if the waveform is correct or not (Fig. 13-13).

Nonsinusoidal Waveshapes

Waveforms which are not sine waves are called **nonsinusoidal.** Many are triangular, rectangular, or square in shape. Because of their nonsinusoidal shapes, two or more amplitude

Fig. 13-13. An oscilloscope.

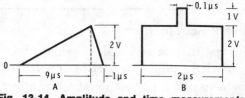

Fig. 13-14. Amplitude and time measurements.

or time dimensions are required to describe them properly. Examples are shown in Fig. 13-14. With such information, it can be determined if the waveform going into or coming out of a circuit is correct.

Circuit Application

To show how waveform representations can be applied to the analysis of a circuit, a sample application is given in Fig. 13-15.

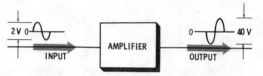

Fig. 13-15. Waveform application.

In the circuit in Fig. 13-15, the amplifier has a sine-wave input and output. Because of the characteristics of this amplifier, the output waveform is a reversal of the input. The output is also greater in amplitude: 40 volts output as compared to 2 volts at the input.

Q13-40. The test instrument which displays waveforms in a circuit is called a(an) _____.

Q13-41. The areas of a circuit to which the test clips are applied are called _____ _____.

Q13-42. A nonsinusoidal waveform is one which is not a _____ _____.

Q13-43. An oscilloscope permits the _____ and _____ characteristics of a waveform to be observed and evaluated.

Q13-44. What is the frequency of the waveform in part A of the illustration at the top of the page?

Q13-45. What is the gain of the amplifier in Fig. 13-15?

Try another circuit (Fig. 13-16).

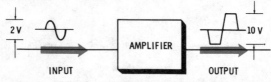

Fig. 13-16. Output waveform different from input.

Instead of a sine-wave output, the tops of both peaks have been flattened. A section of circuit components (different from those in the preceding circuit) is responsible for the different outputs. The amount of gain is also different.

Outputs of two circuits are often fed into a single circuit (Fig. 13-17).

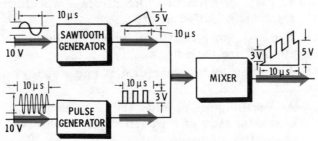

Fig. 13-17. Mixer operation.

The sawtooth and pulse generator have sine-wave inputs of different frequencies. The input frequency of the pulse gen-

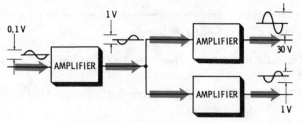

Fig. 13-18. One circuit feeding two.

erator is six times the frequency of the sawtooth generator input. The sawtooth output (its name comes from its shape) has a time duration of 10 μs during which three pulses are produced by the pulse generator. Both outputs are fed to the input of the mixer.

One circuit feeding into two circuits is the reverse of the preceding example. One combination might look like Fig. 13-18.

A low-amplitude sine wave is amplified by the first amplifier and fed to two others. Their outputs are quite different.

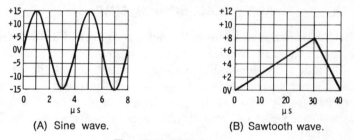

(A) Sine wave. (B) Sawtooth wave.

Fig. 13-19. Waveforms.

Q13-46. What is the frequency of the sine wave in the illustration in Fig. 13-19A?

Q13-47. What is the frequency of the sawtooth wave?

Q13-48. What is the peak-to-peak voltage of the sine wave?

Q13-49. How long does it take the sawtooth wave to rise to +8 volts?

Q13-50. How long does it take the sawtooth wave to decrease back to zero volts?

Q13-51. What is the time duration of a half cycle of the sine wave?

Q13-52. A(an) _____ circuit is capable of superimposing one waveform upon another.

KEEP ELECTRONICS SIMPLE

In learning the electronic principles involved in a variety of circuits, a student technician often gets lost in the many details which are included. You have often heard the expression, "He can't see the forest because of the trees." A person so accused is so involved in examining the details of a single tree he loses sight of how it fits into the entire forest. This analogy aptly fits the study of electronics.

A circuit consists of components which have so many details it is easy for the student to get lost. The usual explanation of how a circuit works often causes a reader to look at a circuit as a group of isolated components.

The explanation must dwell on each of these components to define their place and purpose in the circuit. If the description continues for sentences or paragraphs, the reader has lost the important thread of information that runs through the circuit—how each of the individual components causes the input to be changed into the desired output. The solution to this problem is to keep the "big picture" of what the circuit is supposed to do as a mental image. Then, as each component is discussed, fit its function or action into the appropriate place in the picture. If you make this a habit, you will not get lost in details.

The next two chapters will explain how radio and television receivers operate. Each of the circuits will be described to show how they participate in fulfilling the electronic function of the equipment. Having a picture of how circuits work within familiar equipment will help you understand the details of electronic principles as you continue your study of this subject.

WHAT YOU HAVE LEARNED

1. There is little difference between electrical and electronic circuits. Both are based on identical principles. There is a difference, however, in the manner in which the circuits are applied.

2. Whether it be an application of electricity or one of electronics, circuits operate in a manner that is determined by the effect the components of the circuit have on current that is passing through the circuit or on the voltage that is applied to the circuit.

3. All electrical components, regardless of their name or description, are either resistors, capacitors, inductors, or a combination of these. The effect they have on voltage or current is called resistance, capacitance, and inductance. Since these are the only elements that constitute any circuit, learning their principles well and then applying their effect on current and voltage makes analysis of how a circuit works relatively easy.

4. The changing characteristics of voltage and current can be revealed in their waveforms. A waveform has two dimensions—amplitude (volts or amperes) for height, and time (seconds, milliseconds, or microseconds) for width.

5. Many students learning electronics do well because they retain a mental image of the entire circuit. Then the design *function* of each component can be related to its purpose in producing the overall result. Other students become lost in descriptive detail and forget the designed purpose of the entire circuit.

14

Radio Transmitters
and Receivers

what you
will learn

When you have finished
this chapter you will have
learned what the electro-
magnetic frequency spec-
trum is, what a radio
transmitter is, how it develops a broadcast signal, and
how radio signals are transmitted through the atmosphere. You will also learn how a broadcast signal is
received, and how a radio receiver converts it into sound.
In addition, you will become acquainted with the difference between amplitude and frequency modulation.

In this and the following chapter you will become
familiar with the general principles of operation for
certain equipment. As pointed out previously, an understanding of how electronic equipment works will help
you put descriptions of components and circuits into
proper frames of reference so their meaning is not lost.

ELECTROMAGNETIC RADIATIONS

Energy that radiates from a source is said to be an **electromagnetic wave.** Gamma rays, which are given off by radio-
active particles such as radium, uranium, or atomic-bomb
fragments, are electromagnetic waves. Cosmic rays from the
sun travel extensive distances to the earth as electromagnetic
waves. Electromagnetic waves, which include light, radiated
heat, and radio signals, travel through space at the rate of
300,000 kilometers per second (186,000 miles per second).

Electromagnetic Frequency Spectrum

Electromagnetic radiations differ from each other in terms of their frequencies (stated in hertz). As you recall from the last chapter, the frequency of one of these radiations is the number of times a single cycle repeats itself in 1 second. An **electromagnetic spectrum chart,** showing the relationship of these frequencies, is given in Fig. 14-1.

The chart shows that cosmic rays are radiated at a frequency of around 10^{22} hertz. (The number 10^{22} is 1 followed by 22 zeros, or ten thousand, million, million, million hertz.) At the lower end of the radio portion, radiation frequency is under 10^4, or ten thousand hertz.

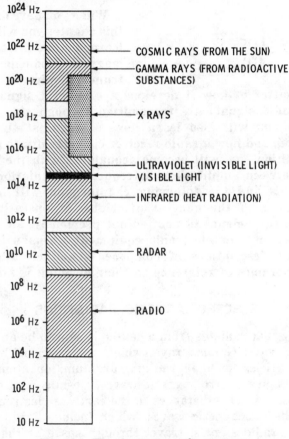

Fig. 14-1. The electromagnetic spectrum.

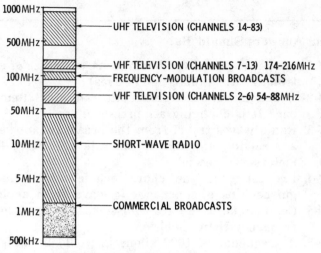

Fig. 14-2. The radio-frequency spectrum.

Assigned Broadcast Frequencies

The Federal Communications Commission (FCC) has assigned specific groups of frequencies to different types of communications transmissions. This is shown in an expansion of the radio frequency portion of the spectrum (Fig. 14-2).

Commercial transmitters (radio and television, for example) are assigned a transmitting frequency in the appropriate part of the radio-frequency spectrum. Transmitters broadcasting in the home radio band, 535 kHz to 1605 kHz (kilohertz), are required by law to be on their assigned frequency within plus or minus 20 hertz.

Q14-1. Cosmic rays and radio waves are examples of _____ _____.

Q14-2. Sound (is, is not) electromagnetic radiation.

Q14-3. Radio waves travel from the broadcast station to a receiving antenna at the rate of _____ kilometers per second.

Q14-4. _____ is the characteristic which distinguishes one electromagnetic wave from another.

Q14-5. Commercial radio transmissions are at a (higher, lower) frequency than television.

Q14-6. A frequency of 1000 kilohertz would be assigned to (commercial, short-wave) radio.

RADIO TRANSMITTERS

The dial on your home radio receiver is marked off in numbers, probably from 550 to 1600 kilohertz (or 55 to 160). By rotating the tuning dial, you select the desired station. Since each local station broadcasts at a different frequency, you are able to select the one you desire. The dial setting indicates the broadcast, or **carrier,** frequency of the station.

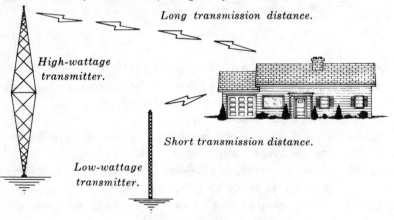

High-wattage transmitter.

Long transmission distance.

Low-wattage transmitter.

Short transmission distance.

Fig. 14-3. Transmission power.

Transmitting Power

You have also noted that some stations come in stronger than others. The stronger stations broadcast at higher power (measured in **watts** or **kilowatts**) than the weaker. Or, if one of the stations broadcasting at equal power is stronger than the other, the stronger station is closer to your home.

The illustration in Fig. 14-3 shows two antennas transmitting at different frequencies in the broadcast band. The one farther away is broadcasting at many kilowatts of power and is able to reach the receiver in the home. The low-wattage transmitter, although nearer, does not have enough power to span the distance. This may explain why you cannot pick up some stations in your general area.

Carrier and Audio Frequency

The frequency assigned to a broadcast station is called its **carrier frequency**. The transmitter and its antenna are designed and tuned to that specific frequency. As its name implies, the carrier frequency carries the reproduction of the sound originating in the studio. Actually, there are two frequencies that leave the transmitter, a **radio frequency** (carrier) and an **audio frequency** (sound). Audio frequencies are classified as being between 20 and 20,000 hertz. The frequency range of most human ears, however, is usually no higher than 15,000 Hz.

Q14-7. A home radio receiver (can, cannot) be tuned to 1 megahertz.

Q14-8. 900,000 hertz (could, could not) be a carrier frequency of a commercial broadcast station.

Q14-9. The power of Station A is one megawatt. Station B is broadcasting at 500 kilowatts. Which station will transmit the longer distance?

Q14-10. Two broadcast stations are equally distant from your home. Assuming your receiver is good, what would be the reason you could not receive one of them?

Q14-11. A human ear (can, cannot) hear a radio frequency.

Q14-12. A frequency of 600 kilohertz is classified as a (an) (audio, radio) frequency.

A Basic Transmitter

The diagram in Fig. 14-4 shows a **functional block diagram** of a typical broadcast transmitter. It is called a functional block diagram because each block is representative of a general electronic function and may include several circuits.

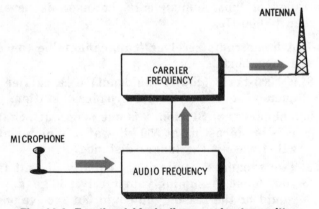

Fig. 14-4. Functional block diagram of a transmitter.

Signal flow direction is indicated by the arrowheads. You can probably already read what the diagram reveals.

Sound enters the **microphone** and is fed to the audio-frequency (af) section. The sound signal, because it is too weak for transmission purposes, is amplified (signal amplitude is increased) and then passed to the carrier-frequency section.

Carrier Frequency—The specific radio frequency (rf) assigned to the broadcast station is developed in the carrier-frequency block. Passing through several circuits, the rf signal is boosted in power (increased in amplitude) to the rated wattage output of the transmitter. Just before the rf carrier is fed to the antenna, the af signal is superimposed on it. Waveforms developed in each block are shown in Fig. 14-5.

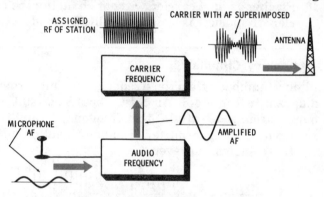

Fig. 14-5. Transmitter waveforms.

Superimposing the Sound—The process of superimposing af on the carrier, as shown in this particular example, is called **amplitude modulation** (am). In amplitude modulation the audio frequency (varying at the changing rate of the original sound) is mixed with the carrier (a constant frequency) in a manner that causes that carrier **amplitude** to vary at the same rate as the audio. The carrier **frequency** remains unchanged.

Q14-13. The drawing in Fig. 14-4 is called a(an) _____ _____ diagram.

Q14-14. Sound enters the af section by way of a device called a(an) _____.

Q14-15. _____ on a block diagram show the signal direction between blocks.

Q14-16. Placing af on a carrier without changing the carrier frequency is called _____.

Carrier-Frequency Circuits

A minimum number of carrier-frequency circuits are shown in the diagram in Fig. 14-6. An actual broadcast station has many more circuits to attain the frequency stability and power required of its transmitter. The additional circuits are similar to those shown, however.

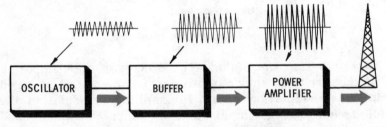

Fig. 14-6. Carrier-frequency circuits.

The Oscillator—The purpose of the **oscillator** is to generate a stable rf signal. The resistance, inductance, and capacitance that make up its input circuit are such that they will not allow the vacuum tube in the oscillator to amplify any other signal but that of the desired frequency. The stable-frequency, low-amplitude output of the oscillator is shown above.

The Buffer—This stage (another name for circuit) is sometimes called an **intermediate power amplifier**, or **frequency multiplier**. In most transmitters it performs three functions. As a **buffer**, the stage isolates the oscillator from the effects of the other circuits. Without this isolation, stray signals may be fed back to the oscillator, causing it to operate at the wrong frequency. As an **amplifier,** the buffer increases the amplitude of the oscillator signal to a level that is between the desired transmitter output and the amplitude of the oscillator signal. In many transmitters the buffer circuit **doubles** (or even **triples**) the frequency of the oscillator output. The oscillator may not be capable of generating the assigned frequency, a transmitter may require several multiplier stages.

The Power Amplifier—The purpose of the **power amplifier** is to increase the amplitude of the rf signal to the power (wattage) requirements of the station. Several stages of power amplification may be required to achieve this. Normally, the audio signal from the af circuitry is fed to the final power amplifier and used to modulate the carrier (Fig. 14-7).

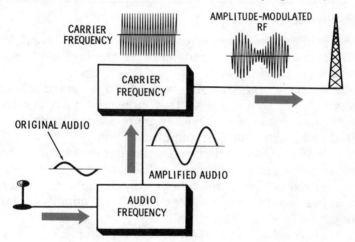

Fig. 14-7. Mixing audio and radio frequencies.

Q14-17. A transmitter circuit which amplifies a signal and increases its frequency is called a(an) _____.

Q14-18. A(an) _____ generates a signal which has a uniform frequency.

Q14-19. _____ amplifier output is measured in watts.

Q14-20. The af and rf are mixed in what stage?

Q14-21. The carrier arrives at the antenna with its waveform (amplitude, frequency) modulated.

Audio-Frequency Circuits

The Microphone—Regardless of the many different types of microphones that are available, even the best develop only a weak signal (Fig. 14-8).

The Audio Amplifier—Although a single stage of audio amplification is sometimes all that is necessary, larger transmitters may have two, three, or more stages to obtain the desired undistorted level of amplitude.

The Driver—Like most circuits, the **driver** obtains its name from its purpose. The driver amplifies the af to the voltage level required to "drive" the tubes of the modulator. The modulator tubes require large changes in signal amplitude to operate properly.

The Modulator—The **modulator** is a power amplifier quite similar to the final circuit of the carrier-frequency block. It amplifies the audio signal to a power level suitable for modulating the carrier power in the final power amplifier. Power output of the modulator is fairly close to half the power of the final carrier amplifier.

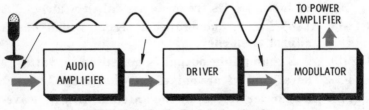

Fig. 14-8. Audio-frequency circuits.

Antennas

If all circuits are operating properly, an am (amplitude-modulated) carrier is fed to the antenna and transmitted into the atmosphere (Fig. 14-9).

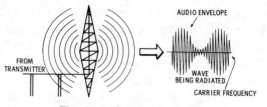

FROM TRANSMITTER

AUDIO ENVELOPE

WAVE BEING RADIATED

CARRIER FREQUENCY

Fig. 14-9. Antenna radiation.

Power is fed to the antenna in the form of both current and voltage. Voltage sets up an electric field along the length of the antenna. Current, in traveling through the antenna (a conductor), sets up a corresponding magnetic field. Both fields vary at the rate of the carrier frequency and at the amplitude and frequency of its audio envelope.

Both fields expand outward and collapse back to the antenna at the rate of the carrier frequency. The outermost waves continue through space and do not return to the antenna. This action is similar to dropping a pebble in a pool. The energy of the waves moves outward in everwidening circles; the water, however, remains in place.

Q14-22. The weak output of a microphone is fed to one or more stages of _____ amplification.

Q14-23. The output of even the best microphones (can, cannot) be fed directly to the modulator.

Q14-24. The output of the _____ is connected to the carrier power amplifier.

Q14-25. For proper modulation, the output of the modulator stage must be _____ that of the power amplifier.

Q14-26. Carrier voltage develops a(an) _____ field and carrier current develops a(an) _____ field on the antenna.

Q14-27. All of the energy in the antenna fields (does, does not) leave the antenna.

Your Answers Should Be:

A14-22. The weak output of a microphone is fed to one or more stages of **audio** amplification.

A14-23. The output of even the best microphones **cannot** be fed directly to the modulator. (Even the most powerful microphones develop a signal that is much too weak to drive the modulator.)

A14-24. The output of the **modulator** is connected to the carrier power amplifier.

A14-25. For proper modulation, the output of the modulator stage must be **half** that of the power amplifier.

A14-26. Carrier voltage develops an **electric** field and carrier current develops a **magnetic** field on the antenna.

A14-27. All of the energy in the antenna fields **does not** leave the antenna. (Only the outermost waves.)

A RADIO RECEIVER

The block diagram for a radio receiver similar to the one in your home is shown in Fig. 14-10.

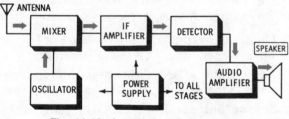

Fig. 14-10. A typical radio receiver.

The purpose of the radio receiver is to convert the amplitude modulation on the carrier back to its original sound. As the carrier increases in ever-widening circles on leaving the transmitter antenna—like ripples in a pool—its energy decreases in amplitude. The increasing circumference of the circles causes power in the waveform to be distributed over an ever-increasing area. By the time the signal reaches the receiver antenna it is rather weak, usually around a few

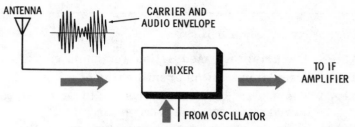

ANTENNA CARRIER AND AUDIO ENVELOPE

MIXER

TO IF AMPLIFIER

FROM OSCILLATOR

Fig. 14-11. Antenna and mixer.

thousandths or millionths of a volt. The receiver, therefore, must amplify the received signal to a level that will operate the speaker within the hearing range of the human ear. The receiver must also extract the audio component (the **envelope**) from the carrier. The carrier brings the signal to the receiver, but has no value in the reproduction of the audio frequency in the receiver.

Receiver Circuits

The Power Supply—Each receiver has a power supply. Its purpose is to convert 117 volts ac from an electrical outlet (or to provide dc if the receiver is battery-operated) to voltages that will operate the receiver properly.

The Antenna and Mixer—Carrier frequencies from all stations within range of a receiver appear on the antenna of the receiver. When you turn the dial of your radio to a specific station, you adjust the electronic components of the **mixer** input so that the receiver will accept a particular carrier frequency and reject all others. The received carrier enters the mixer to be amplified. Some radios have, in addition, an **rf amplifier** between the mixer and antenna (Fig. 14-11).

Q14-28. What part of the received radio wave does the receiver convert back into original sound?

Q14-29. A radio wave decreases in power as the circumference of its wave increases. What is the approximate amount of voltage that enters the receiver antenna?

Q14-30. The _____ _____ converts ac to voltages required to operate the receiver circuits.

Q14-31. A single broadcast frequency appears at the input of the (antenna, mixer).

The Oscillator—The receiver oscillator is similar to its counterpart in the transmitter. Both generate a signal of constant frequency and amplitude. The purpose of the receiver oscillator is slightly different, however. It is designed to generate a frequency that is a constant number of kilohertz above the carrier frequency, regardless of the station to which the receiver is tuned. The tuning dial changes the values of the electronic components in the frequency-generating circuit of the oscillator at the same time it is adjusting the frequency-reception components of the mixer. The arrangement is such that the oscillator will always be tuned 456 kilohertz (or a similar frequency) above the frequency of the carrier being accepted by the mixer. The output of the oscillator is fed to the mixer, as shown in Fig. 14-12.

The Mixer—The carrier and oscillator frequencies combine in the mixer tube and four different frequencies appear at the output. One of these four is the **difference** between the oscillator and the carrier frequencies, and is usually 456 kilohertz. The other three are rejected by the next stage. The mixer and oscillator together are called the converter.

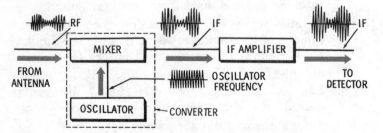

Fig. 14-12. Mixer, oscillator, and if amplifier.

The IF Amplifier—The abbreviation for intermediate frequency is **if**. In most home receivers the if is 455 or 456 kHz. Amplifying a single frequency in the if circuit is much easier and causes less distortion than if it were necessary to tune this amplifier to each of the many station frequencies. The only purpose of this stage is to amplify the if (which still retains the original audio frequency) and pass it on to the detector.

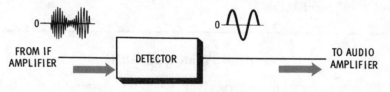

FROM IF AMPLIFIER → DETECTOR → TO AUDIO AMPLIFIER

Fig. 14-13. Detector.

The Detector—The purpose of the **detector** is to remove the audio component from the if waveform. The audio envelope is the same (although reversed) at the top of the waveform as it is at the bottom. The detector circuit is so designed that it accepts only the audio frequency at the top and rejects the if frequency in the waveform.

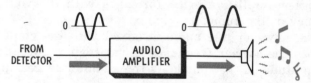

FROM DETECTOR → AUDIO AMPLIFIER →

Fig. 14-14. Audio amplifier and speaker.

The Audio Amplifier—The final circuit in the receiver amplifies the af fed to it by the detector. The amount of amplification can be varied by the volume-control knob on the front of the receiver. The output of the audio amplifier is applied to the speaker voice coil, causing the speaker cone to reproduce the sound that originated at the studio.

Q14-32. The _____ removes the af from the if waveform.

Q14-33. The oscillator develops a signal at a constant _____ and _____.

Q14-34. A converter combines the functions of _____ and _____.

FREQUENCY MODULATION

The transmitter and receiver with which you have just become familiar employs amplitude modulation (am) to carry the audio. Another method of superimposing audio on a carrier is called **frequency modulation** (fm). Its process is quite different. The two are compared in Fig. 14-15.

Both am and fm start out with a carrier frequency and an audio frequency (sound originating in the studio). In amplitude modulation, as you already know, the sound is superimposed on the carrier frequency (which is constant) by varying the carrier **amplitude** in conformance with the voltage and frequency of the audio.

In fm, however, the audio is mixed with the rf in such a way that the carrier **frequency** is varied in accordance with the amplitude of the sound. As the audio cycle goes positive,

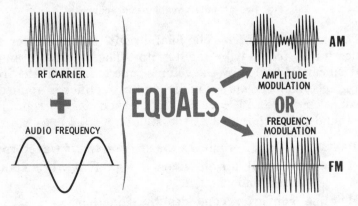

Fig. 14-15. Comparing am and fm.

carrier frequency increases. When the audio cycle goes negative, carrier frequency decreases. The sum of the two changed frequencies in one audio cycle is still equal to the original carrier frequency.

One of the advantages of frequency modulation is its freedom from distortion. Noise and other forms of distorting voltages in the atmosphere or receiver are added to amplitude modulation. Since fm does not depend on a changing amplitude to carry audio, noise has little or no effect on it. This is part of the reason for the clarity of sound that you get from an fm receiver (Fig. 14-16).

Fig. 14-16. Noise has no effect on fm.

Q14-35. In am, the carrier _____ changes to match the audio.

Q14-36. In fm, the carrier _____ changes to match the audio.

Q14-37. An fm receiver is (more, less) subject to atmospheric noise than an am receiver.

WHAT YOU HAVE LEARNED

1. Radiant energy is given off by electromagnetic waves. The electromagnetic spectrum includes cosmic rays, X rays, visible and invisible light, infrared, radar, as well as radio waves.

2. A radio transmitter is a device that produces electromagnetic waves in the radio portion of the spectrum. Its essential functions are the development and amplification of a carrier frequency and modulating it with an amplified audio frequency. A specific carrier frequency is assigned to each radio station. The distance that the carrier, with its superimposed audio, travels is deter-

mined by the power that is developed in the final stage of the transmitter.

3. Energy in the form of voltage and current is fed from the transmitter to an antenna. This sets up electric and magnetic fields around the antenna that expand and collapse at the frequency of the carrier. Part of the energy is in the form of electromagnetic radiations and is transmitted through the atmosphere. The farther it travels, the weaker the signal becomes.

4. All carrier signals within range are picked up by the receiver antenna. The tuning control on the front of the receiver adjusts the input of the mixer so that only the desired station carier frequency is received. At the same time, it adjusted an oscillator to generate an if above the carrier frequency. Carrier and oscillator frequencies are joined in the mixer and the difference between the two, the intermediate frequency, is amplified and fed to the if amplifier. Here the signal and its audio components are further amplified. The next stage (detector) extracts the audio component and passes it to the final stage (audio amplifier). The audio is amplified and fed to the speaker, causing the cone to reproduce the sound that originated at the studio.

5. Amplitude (am) and frequency (fm) modulation are two methods of transmitting audio on a carrier. When am is used, the amplitude of the carrier varies accord- to the loudness (amplitude) and frequency of the audio. In fm, the frequency of the carrier is varied instead of the amplitude. Atmospheric and receiver noises are less bother to fm transmissions.

15

Television Transmitters
and Receivers

what you will learn

In this chapter you will learn how a television transmitter develops both picture and sound signals. You will gain more knowledge about antennas and the problems of sending electromagnetic waves through the atmosphere. You will also become familiar with how a television receiver converts electronic signals into picture and sound reproductions. Learning the basic principles of television transmitters and receivers is no more difficult than learning the principles of radio. The basic electronic principles are the same. You will find this to be true for any electronic equipment, regardless of how complicated it may seem.

THE TELEVISION TRANSMITTER

There is actually very little difference between radio and television transmitters. As you recall, the functional block diagram of a radio transmitter contains a microphone, an audio-frequency section, a carrier-frequency section, and an antenna. A television transmitter has more operations to perform than a radio transmitter. Its functional block diagram, therefore, contains more sections (Fig. 15-1). The functions of the two transmitters are quite similar, however.

Functional Block

The radio transmitter has the single problem of putting sound on a carrier. The tv transmitter must modulate two carriers, one with sound and the other with **video** (picture).

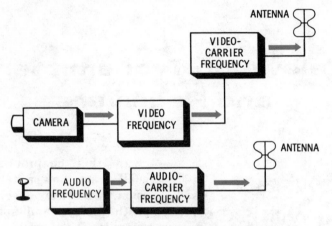

Fig. 15-1. A tv-transmitter functional block diagram.

Video-Frequency Functions—The difference between the functional blocks of a radio transmitter and a tv transmitter is the addition of a **video-frequency** function in the tv transmitter. The audio-frequency section is in a separate channel of its own. Shown in the tv block diagram in Fig. 15-1 is a camera which sends a weak picture signal to the video-frequency section to be amplified. The output of this section is a **video** frequency (higher than audio) used to modulate a **very high frequency** (vhf) generated in the carrier block. Superimposing the video (picture) on the carrier is done by amplitude modulation, the same process used in an am radio transmitter.

Audio-Frequency Functions—A microphone feeding a signal to the audio-frequency section is shown at the bottom of the illustration above. The sound signal from this microphone is amplified and used to frequency-modulate a separate carrier. This modulated carrier is then fed to an antenna. In effect, there are two transmitters for tv—one for transmitting the picture and the other for transmitting the sound. In practice, a single antenna is usually used to transmit both carriers.

THE TV AUDIO TRANSMITTER

As previously stated, a tv audio transmitter uses the frequency modulation method. In the preceding chapter, you learned that fm is a process in which the frequency of a

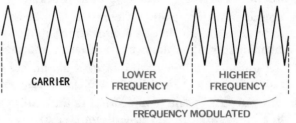

Fig. 15-2. Frequency modulated wave.

carrier is varied in accordance with the amplitude of an audio signal.

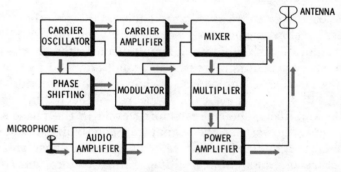

Fig. 15-3. Block diagram of an fm transmitter.

Q15-1. The sections of a functional block diagram of a radio transmitter are a microphone, _____ _____, _____ _____, and antenna.

Q15-2. Sound in an am radio transmitter is placed on the carrier as a(an) _____ _____; in a television transmitter, it is done by _____ _____.

Q15-3. An audio frequency modulates a sound carrier; a(an) _____ frequency modulates a picture carrier.

Q15-4. The outputs of the _____ _____ and _____ _____ blocks of the sound transmitter are fed to the modulator.

Audio Modulation of the Carrier

The oscillator in an fm transmitter, as in any other transmitter, develops a constant frequency at a uniform amplitude. The output of the oscillator is simultaneously fed to a carrier amplifier (where it is increased in amplitude) and to a phase-shifting circuit. A single cycle is shown in the oscillator block in Fig. 15-4. The same cycle appears in the output of the carrier amplifier. The corresponding cycle in the phase-shifting circuit shows that the signal has been shifted (moved) to the right a quarter of a cycle. This is called **phase shifting.** The starting, maximum positive, return-to-zero, maximum negative, and ending points occur one quarter of a cycle later in the lower block than they do in the upper block. Since there are 360° (one way of designating the period of a sine-wave cycle) in a complete cycle, the lower waveform has been shifted in phase by 90°.

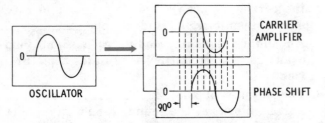

Fig. 15-4. Phase shifting.

Modulator—Amplified sound signals from the audio amplifier and the phase-shifted carrier frequency meet in the modulator. The result of the meeting is a modulated output with an amplitude that varies in accordance with the amplitude of the audio, but still phase-shifted a quarter cycle.

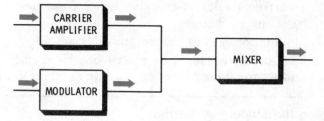

Fig. 15-5. Phase-shifted audio and carrier joined in mixer.

Mixer—This circuit mixes the outputs of the carrier amplifier and the modulator to produce a variable frequency. This new signal is the carrier frequency changed by an amount determined by the amplitude variations of the audio. If the two inputs to the mixer had been in phase, this frequency variation of the carrier could not have occurred. Since equivalent points of the two waveforms are changing at different times, the audio variations of the modulator signal either add cycles to or subtract cycles from the constant frequency of the carrier. When the audio content of the modulator frequency goes positive, it causes a corresponding increase in frequency of the carrier. When the audio content goes negative in its cycle, the carrier frequency decreases a proportionate amount. The output of the mixer then is the basic carrier changing in frequency in accordance with the amplitude of the original sound.

Q15-5. The frequency of any transmitter is generated in a(an) _____ circuit.

Q15-6. The phase-shifting circuit changes the phase (starting point) of the sine wave a(an) _____ of a cycle, or _____ degrees.

Q15-7. Modulation occurs (before, after) power is amplified.

Q15-8. The audio frequency is placed on the carrier by _____ to and _____ from the carrier cycles.

Amplifying the Modulated Carrier

Multiplier—The purpose of the multiplier stage is to amplify and increase the frequency of the modulated carrier. In some transmitters, several such circuits may be required. In addition to raising the amplitude of the carrier to that which is required for transmission purposes, multiplying the frequency also makes the modulated variations more pro-

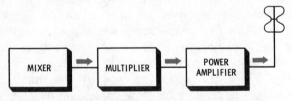

Fig. 15-6. Mixer, multiplier, and power amplifier.

nounced. Those portions of the audio signal having higher amplitudes cause a wider variation of the carrier frequency than do those portions having lower amplitudes (Fig. 15-6).

Amplifier—You will note that the carrier is modulated prior to the power amplifier stage. In amplitude modulation it is necessary to superimpose the audio in the final power stage, which is usually at very high power. Since the final amplifier of the audio-frequency sections must have a power output that is close to 50% of that of the carrier amplifier, tubes and other parts must be large and expensive. In fm, the entire modulated carrier is raised to the correct power level in a single stage, making the fm transmitter more economical in this respect than its am counterpart. The output of the power amplifier goes directly to the antenna.

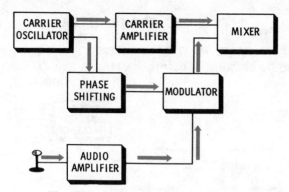

Fig. 15-7. Audio added to the carrier.

Audio Transmitter Review

Sound from a television studio is added to the carrier wave by frequency modulation. The block diagram in Fig. 15-7 shows the transmitter portion which mixes the two frequencies.

The **oscillator** generates a stable frequency which is simultaneously fed to an **amplifier** and a **phase-shifting** network. The phase-shifting stage shifts the frequency by a quarter of a cycle. The carrier and the shifted signal are no longer in step. The amplified audio from the microphone is mixed with the out-of-phase signal in the **modulator.** The output of the modulator is a series of sine waves that vary in amplitude in accordance with the amplitude of the original sound.

The outputs of the carrier amplifier and the modulator combine in the **mixer.** The output of the mixer is a signal that varies in **frequency** according to the **amplitude** of the modulating signal. The fm signal is then multiplied in frequency several times and increased in power by the stages shown in Fig. 15-8.

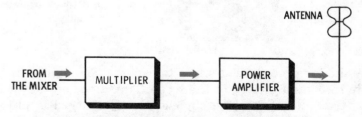

Fig. 15-8. The modulated carrier is fed to the antenna.

TV VIDEO TRANSMITTER

A brief summary of what happens in the tv video transmitter might be helpful before its stages are discussed.

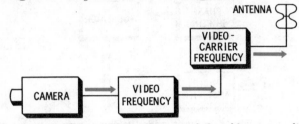

Fig. 15-9. Functional block diagram of the video transmitter.

In the video-frequency section, video signals from the camera are amplified and fed to the final power amplifier of the carrier-frequency section (Fig. 15-9). Here, the carrier is amplitude-modulated by the video.

Camera—There are several different types of tv cameras. The **iconoscope, image dissector,** and **image orthicon** are examples. The latter is the type most frequently used in tv broadcasts. Although the manner in which they accomplish their purposes differs, their basic operating principles are the same (Fig. 15-10).

The camera, much like its photographic counterpart, deposits a scene through a lens on a plate within the camera. Light rays from all parts of the scene are focused through the lens, reproducing the image on the plate. If the plate were a photographic negative, the light rays would excite deposits of light-sensitive materials in proportion to the intensity of light, varying from white through shades of gray to black.

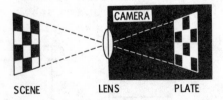

Fig. 15-10. Camera operating principle.

SCENE   LENS   PLATE

A similar process occurs in a tv camera. The light-sensitive plate receives a picture of the scene. Tiny areas on the chemically treated plate are thereby electrically charged in proportion to the light intensity of that part of the scene.

Scanning

A very narrow beam of electrons is moved back and forth across the plate from top to bottom. The beam samples the intensity of the charge in each of the tiny areas. The amount of each charge indicates whether that portion of the scene is black, white, or some shade of gray.

525 LINES OF SCAN

SCANNED 30 TIMES EACH SECOND

(A) Camera plate.

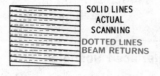

SOLID LINES ACTUAL SCANNING

DOTTED LINES BEAM RETURNS

(B) Portion of plate showing scan lines.

Fig. 15-11. Electron beam scanning.

As shown in Fig. 15-11A, the plate is **scanned** (movement of electron beam) in a sequence of 525 lines from top to bottom. A complete scan of the plate (525 lines each time) is made 30 times each second. The same procedure is duplicated on the screen of your receiver. In a tv receiver with a 17-inch screen, the electron beam in the picture tube travels across the screen at approximately 13,000 miles per hour.

Part B shows how this scanning is accomplished. The beam moves across the plate in the camera from left to right, sampling the intensity of each tiny area it passes. At the end of the line the beam is **blanked** (shut off) and returned to the left side of the plate to start the next line. The beam is turned on again and samples the second line. This process is continued until the bottom of the plate is reached. The beam is blanked and returned to the upper left-hand corner to start scanning again. When the beam is on and moving from left to right sampling the intensity on the plate, it is said to be **scanning**. When it is shut off and being returned to a new starting point, it is **retracing**.

Q15-9. The video is placed on the carrier by _____ _____.

Q15-10. The image plate is _____ by an electron beam.

Q15-11. How many lines does the beam trace each second?

Q15-12. How many times a second is the beam blanked?

Interlaced Scanning

Because of problems in controlling the beam and of noticeable flicker to the viewer when **line-by-line scanning** is performed, the beam is caused to scan every other line.

As Fig. 15-12 shows, the first scan starts at line 1, samples the charged areas, and is retraced to line 3. This action continues to the bottom of the plate, scanning the odd-numbered lines. When it reaches the bottom, the beam returns to the top of the plate and scans the even-numbered lines. Each scan, top to bottom, requires 1/60 of a second. To scan the entire plate, the beam requires two passes, which takes a total time of 1/30 of a second. On the receiver screen a new image is being presented on every other line 60 times a second, a line-tracing frequency that cannot be noticed by the eye. If it were being done at the rate of 30 times a second, the eye might be able to see the changes, which would be recognized as a flicker. This process of scanning every other line is called **interlaced scanning.** The camera thus identifies the light and dark areas of a scene and converts this information to currents and voltages that change in proportion to the light intensity.

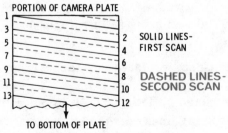

Fig. 15-12. Interlaced scanning.

Timing Generator

The timing of the scanning events is very critical. The beam of electrons must begin at a precise point near the top of the camera plate and scan every odd-numbered line in 1/60 of a second. The electron beam must be blanked out precisely at the end of every line and at the end of the field. A complete scan of all the odd-numbered lines (or even-numbered lines) is called a field.

When the odd-numbered field has been completed, the blanked beam must be returned to a new position at a precise time to begin scanning the even-numbered field. Each action and position of the camera beam must be followed precisely by similar action in your tv receiver at home. The stage in the tv transmitter that establishes this precise timing is known as the **timing generator,** sometimes called the **blanking** or **sychronizing** stage.

The timing generator in Fig. 15-13 feeds pulse waveforms to the camera tube. The amplitude and timing of the pulses are such that they **synchronize** (cause all events to take place at precise time intervals) scanning, blanking, retracing, and positioning of the electron beam. The same timing pulses (for synchronizing the same events in the receiver) are fed, with the amplified video, to another stage of video amplifiers. From this point the entire signal—video and timing pulses— is passed to the final amplifier of the carrier for modulation purposes.

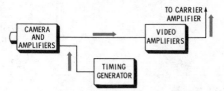

Fig. 15-13. Video-frequency stages.

Q15-13. _____ scanning skips every other line.

Q15-14. By this method, a given line on the receiver screen is scanned _____ times a second.

Q15-15. Scanning is synchronized by a(an) _____ _____.

Q15-16. What is contained in the video-output amplifiers?

Video Modulation

The video and timing pulses are placed on the carrier frequency by amplitude modulation. The process is similar to the method used by am radio stations.

Video Signals—As you have learned from the preceding discussion, a video signal contains a great deal of information. A series of video waveforms is shown in Fig. 15-14. Remember that a waveform contains only two dimensions—amplitude and time.

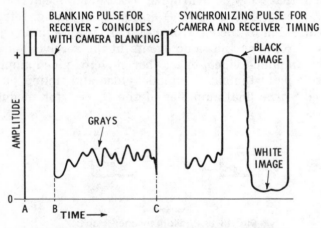

Fig. 15-14. Complete video for two scanned lines.

Carrier Frequency—The carrier-frequency section is similar to the same circuits in a broadcast radio transmitter. (See the diagram in Fig. 15-15.)

The oscillator generates a continuous and constant frequency. The output of the oscillator is increased in frequency and amplitude by the multiplier and amplifier sections. In the

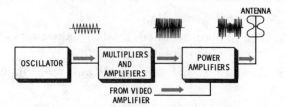

Fig. 15-15. Television carrier frequency.

power amplifier, the carrier is raised to the desired power level required by the station, and is amplitude-modulated by the video signal.

For vhf (very high frequency—Channels 2 to 13), the frequency of the carrier is between 54 and 216 megahertz. For uhf (ultra high frequency—Channels 14 to 83), the carrier is between 470 and 890 megahertz. Transmission of signals at these frequencies is quite different from that for the lower radio frequencies. High frequencies have short wavelengths. A **wavelength** is the time duration, or length, of one cycle. The higher the frequency of a signal, the shorter is its wavelength.

Q15-17. The image scanned by the camera is changed into a(an) _____ frequency.

Q15-18. A video signal has _____ and _____ dimensions.

Q15-19. In the diagram in Fig. 15-14, a full video cycle is from (A to B, B to C, A to C).

Q15-20. The beginning of a scanned line on the receiver screen coincides with (A, B, C).

Q15-21. Using the same letters, the electron beam is retracing between time _____ and time _____.

Q15-22. A black image appears as a (more, less) positive voltage than a shade of gray.

Q15-23. During retrace time in a video cycle, two pulses appear. What are they?

Q15-24. How many blanking pulses are shown in Fig. 15-14?

Q15-25. Vhf has (shorter, longer) wavelengths than uhf.

Q15-26. A tv video signal is _____ modulated.

TELEVISION TRANSMITTING ANTENNAS

An antenna, as you recall, develops an electromagnetic field around itself when current is passing through it. Current flows back and forth through an antenna in accordance with the rise and fall of the carrier-wavelength frequency and amplitude.

Television Wave Propagation

Since the electrical wavelength of a tv carrier is shorter than that of a radio-broadcast carrier, the length of the tv antenna is correspondingly shorter. There is also a difference in the way short and long wavelengths travel through space.

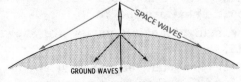

Fig. 15-16. Radiated long waves.

The diagram illustrates the **propagation** (travel of electromagnetic radiation) of long waves (Fig. 15-16) as they radiate from the antenna of a commercial radio station. The frequency is between 535 and 1605 kilohertz. The radiated energy has a **space wave** that travels essentially in a straight line. Low frequencies also have a **ground wave** that hugs the ground until the radiated power decreases so much with distance that reception is no longer possible.

The short wavelengths of a television transmission depend on a different method of wave propagation (Fig. 15-17).

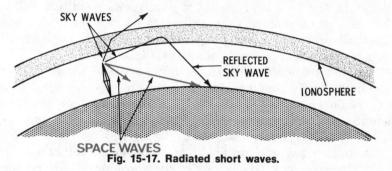

Fig. 15-17. Radiated short waves.

High frequencies radiate a space wave and a **sky wave**. Both of these waves travel essentially in straight lines. To receive a space wave, the receiving antenna must be within line-of-sight of the transmitting antenna. If the receiving antenna is beyond the horizon, the space wave (if it still retains sufficient power) passes over it.

Sky waves, also traveling in a straight line, head out into space. When the sky waves are 50 to 75 miles out, depending on the time of day, they encounter the **ionosphere**, a layer of charged particles that cause the short-wave radiations to bend. Waves that enter the ionosphere at a sharp angle are bent back to earth. At an angle close to 90° (perpendicular), the bending is not enough for the signal to return to earth, so it continues to travel toward higher altitudes. If they are of sufficient power, reflected sky waves can sometimes be picked up by receiving antennas.

Q15-26. Tv frequencies have _____ and _____ waves.

Q15-27. _____waves may be reflected by the ionosphere.

Q15-28. Tv radiations are (short, long) waves.

THE TELEVISION RECEIVER

There are many different models of tv receivers. They differ in the type and numbers of circuits used, as well as in the size of picture tube and style of cabinet. Since they all must process the same signals from a tv transmitter, the function of their circuits must be identical.

The illustration in Fig. 15-18 shows a single antenna bringing both the fm sound carrier and the am video carrier to the rf (radio frequency) circuits. This is satisfactory since the sound- and video-carrier frequencies are fairly close together (the sound carrier is 4.5 megahertz higher).

Both carriers are amplified and converted into an intermediate frequency (if). The if signals are amplified and then separated, each being sent to its proper section. In the sound section, the audio component of the frequency-modulated wave

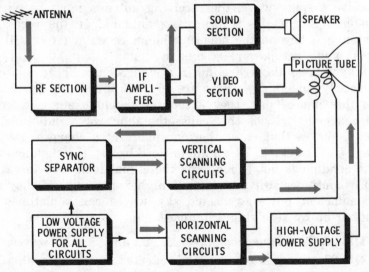

Fig. 15-18. Block diagram of a tv receiver.

is extracted and sent to the speaker. In the video section, the picture signals and blanking pulses are taken from the amplitude-modulated wave and sent to the cathode-ray (picture) tube.

Synchronizing (sync) pulses time the controlling voltages of the vertical and horizontal circuits. Outputs of these stages cause the image to be placed on the screen of the cathode-ray tube.

RF Section

In most tv sets the rf section (sometimes called the **front end** or **tuner**) normally consists of an rf amplifier, a mixer, and an oscillator (Fig. 15-19).

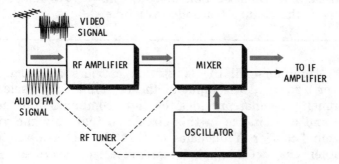

Fig. 15-19. Rf section of a tv receiver.

Dashed lines to two or more circuits in a diagram indicate that a single control has an effect on each circuit or part indicated.

Q15-29. _____ and _____ carriers are received by a single antenna.

Q15-30. The two frequencies differ by _____ megahertz.

Q15-31. The carrier containing the synchronizing pulses is at a (lower, higher) frequency than the audio carrier.

Q15-32. Outputs from the if amplifier are fed to the _____ and _____ sections.

Q15-33. What stages are tuned by the receiver channel selector switch?

Q15-34. What section delivers the picture image to the picture tube?

RF Amplifier—All the tv station carriers reaching your receiver appear at the input of the rf amplifier. In selecting a channel, the tuner provides the right combination of inductance and capacitance in this circuit so that only the video- and sound-carrier frequencies of that channel are amplified. All other channel frequencies are rejected. There are very few receivers that do not have an rf amplifier.

Mixer and Oscillator—The mixer is tuned to the same frequency as the rf amplifier. Its purpose is to develop the intermediate frequency for the if amplifiers located in the sound and video sections. The mixer (converter) does this in the same manner as the mixer in the am radio receiver. The oscillator develops a frequency that is the desired if above the video- and sound-carrier frequencies. The oscillator and carrier frequencies are mixed to produce the if difference frequency at the output of the mixer. This frequency is then fed to the if amplifiers with the appropriate video and sound modulations still existing.

Fine-Tuning Control—Some sets have a **fine-tuning control** in addition to the channel selector. The fine-tuning control adjusts the value of a component (usually a capacitor) in the oscillator circuit. The change in value of this component causes an appropriate change in the frequency of the oscillator, allowing the if for sound and video to be tuned more precisely.

Sound Section

You may have noticed the similarity between the mixer and oscillator of a radio receiver and the corresponding circuits in a tv receiver. In fact, most of the circuits of the sound section are similar in operation to those found in an fm radio (Fig. 15-20).

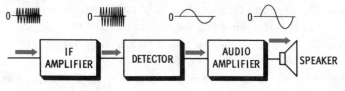

Fig. 15-20. Tv-receiver if and sound sections.

IF Amplifiers—The if signal contains both frequency and amplitude modulation. The if frequency is usually near 45 megahertz, which is a much higher frequency than the 455-kHz if usually found in an am radio. This higher if frequency modulated by both the audio and video signals is more difficult to amplify. Thus, the gain (amount of amplitude increase between circuit input and output) is low, and more than one stage of if amplification is necessary. Depending on the quality of the receiver, the if section will have two, three, or four if amplifiers, one after the other. This row of amplifying circuits is sometimes called the **if strip.**

Q15-35. Front end is another name for the ____ _____.

Q15-36. What are the two carriers that enter the rf amplifier from the antenna?

Q15-37. What are the frequencies that become inputs to the mixer?

Q15-38. The selector switch tunes the mixer to the same input frequency as the _____ _____.

Q15-39. The oscillator is tuned to a (higher, lower) frequency than the video carrier.

Q15-40. The if amplifiers are circuits in both the _____ and _____ sections.

Q15-41. There are (more, fewer) circuits in a tv if strip than in the similar section of a home radio.

Sound Detection and Amplification

Because the tv sound is contained in the form of frequency modulation, the method for removing the audio component is different from that for am.

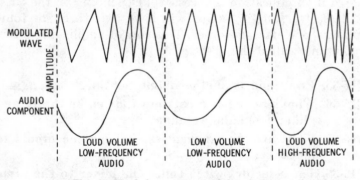

Fig. 15-21. A frequency-modulated wave.

The Detector—Both fm and am detectors have the same purpose—to remove the audio component from the modulated intermediate frequency. For fm, the variations in frequency are changed into voltage variations by the detector. The output of the detector (quite frequently called a **discriminator** because it discriminates between af and if), is an audio frequency representing the tone and amplitude of the sound originating at the studio.

The difference in volume, as shown in Fig. 15-21, is determined by the distance between cycles of the carrier (or if). For a given carrier, its frequency is the same for each cycle of audio. Carrier frequency decreases (cycles farther apart) during the negative portion of an audio wave and increases (cycles close together) during the positive portion. The greater the volume of the audio, the greater is the difference in carrier frequency between the positive and negative half cycles of the audio. A weaker volume shows less difference (carrier frequency will be more uniform throughout the audio cycle) between positive and negative half cycles of the audio.

High and low tones are determined by the number of times the periodic carrier-frequency variations repeat themselves. For low audio tones, repetition of cycles (audio frequency) is less often. For high tones, repetition of cycles occur a greater number of times per second.

The Audio Amplifier—The purpose and method of amplifying audio are identical in am and fm receivers. The amplifier in either system raises the amplitude of the pure audio signal to the level required for operating the speaker. The volume control and tone control (if the tv set has one) are normally separate variable resistances in the first two audio-amplifier stages.

Q15-42. Video detectors are designed to remove _____ modulation from the _____ frequency.

Q15-43. The audio component of an fm signal is removed by a(an) _____ circuit.

Q15-44. In frequency modulation, the distance between cycles of the carrier frequency determines the (volume, tone) of the sound.

Q15-45. A high tone requires (more, less) repetition of the if variations than a low tone.

Q15-46. A discriminator separates the audio modulation frequency from the _____ _____.

Video Section

The purpose of the video section is to detect and amplify the picture signal and distribute its signal components to the correct stages of the set.

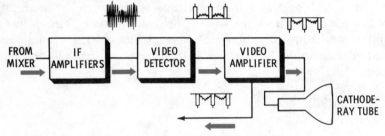

Fig. 15-22. Tv-receiver video section.

IF Amplifiers—The output of the if amplifiers contains the video signal and is fed to the video detector, as shown in Fig. 15-22.

Video Detector—Like the detector used in an am radio, a tv video detector is usually a diode. As you remember from a previous chapter, a diode conducts current in one direction only. When the modulated signal is applied, the detector conducts only during the time the waveform is going positive. The varying frequency representing the sound and the negative portion of the video signal cannot pass through the detector stage. Therefore, the output of the detector is identical to the picture signal as it left the camera and before it was placed on the carrier.

Video Amplifiers—The output of the detector is a relatively weak signal, not strong enough to cause a reproduction of the picture. Video amplifiers are, therefore, required to achieve the necessary signal amplitude. These must be wide-band amplifiers because the frequency content of the picture signal covers a wide frequency range (Fig. 15-23).

The output from the video-amplifier section is the reverse (upside-down) of the waveform that entered its input. This is the condition of the waveform that is desired. If the image and blanking pulses were positive going instead of negative going, the blacks of the image would appear as whites, and the whites as blacks.

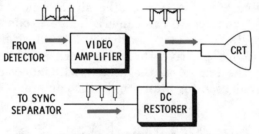

Fig. 15-23. Video amplifier and dc restorer.

Contrast Control—The contrast control is actually a variable resistance in one of the video-amplifier stages. Its purpose is to vary the amount of output from the amplifier with which it is associated. This increases or decreases the amplitude difference between the voltages representing the white and black portions of the image. If the amplifier output is increased, the difference between the two voltages becomes greater and the contrast is increased.

Q15-47. The contrast control changes the amount of _____ of a video amplifier.

Q15-48. The video detector selects the (positive, negative) portion of the picture signal and rejects the _____ frequency.

Q15-49. In the output of the video-amplifier section the sync pulse has a voltage (more, less) negative than the image.

Image Display

In the illustration in Fig. 15-24, zero voltage is shown on the reference line. Any portion of the signal below this line is negative. As you recall, the camera image produces a signal in which whites are more positive than blacks and grays, and grays more positive than the blacks. The cathode-ray tube places the image on the screen with an electron beam similar to that used for scanning in the camera. Video signals

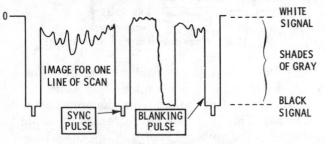

Fig. 15-24. Video waveform content.

fed to the cathode-ray tube control the number of electrons striking the **fluorescent** screen. The fluorescent material on the screen gives off light in proportion to the number of electrons that strike it.

To reproduce blacks, the beam must be shut off. Whites require a maximum number of electrons. The video signal controls the number of electrons by the value of its negative voltage. Negative voltage repels electrons. A highly negative portion of the image signal (black) stops electron flow completely.

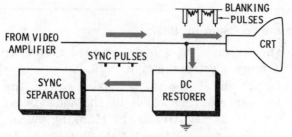

Fig. 15-25. Sync and blanking pulses.

DC Restorer—To achieve proper values of the image and blanking voltages that control the number of electrons in the beam, a definite voltage reference level must be established and maintained. In other words, zero voltage is the reference shown in Fig. 15-25, and must be at the top of the waveform. If the waveform varies above or below this zero reference, the video and blanking will not appear on the screen properly. The circuit in early tv sets that maintains this level is called a **dc restorer.**

Brightness Control—Another front-panel adjustment, the **brightness control,** is a variable resistor in the picture-tube circuit. The purpose of this control is to adjust the position of the waveform on the zero reference level to a point that provides the best screen brightness for viewing purposes. If the control is adjusted so that the near-white amplitudes in the image are brought closer to the zero reference, more electrons strike the screen, making it brighter. When the control is turned in the other direction, so that even the white amplitudes are below the zero reference, fewer electrons strike the screen and the entire picture is darker.

Q15-50. The screen of a cathode-ray tube glows brighter if (more, fewer) electrons strike it.

Q15-51. To put "black" on the picture tube, the voltage representing black in the video waveform must be as (negative, positive) as the _____ pulse.

Q15-52. A zero reference level of the video waveform is established by the _____ _____.

Q15-53. The _____ _____ adjusts the zero reference level for the desired brightness of the screen.

Scanning

In the discussion thus far, video and blanking pulses have been fed to the cathode-ray tube for each scanned line of the picture waveform entering the set. The picture portion of the waveform controls the intensity of the electron beam, while

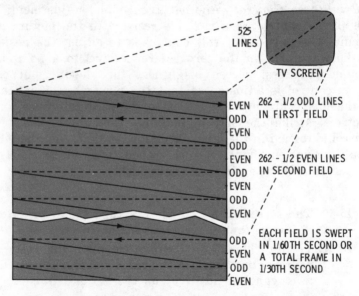

Fig. 15-26. Screen and camera beams are synchronized.

the periodically appearing blanking pulses shut the beam off at the proper intervals (Fig. 15-26).

306

Some method is needed to move the beam on the receiver screen from side to side and top to bottom in synchronization (in step) with the action that takes place in the camera. Each of the video waveforms represents one particular scan line among the 525 lines that should appear on the screen for a complete picture. Up to this point all of the video waveform has been used except the small **sync pulses** that are on top of the blanking pulses.

There are 525 horizontal lines in a complete picture on a tv screen. Each line represents an image line scanned by the tv camera and a screen line to be **swept** (reverse of scan) by the electron beam in the tv tube. The entire 525 lines are called a **frame.**

The camera scans every other line (interlaced scanning) for ease in electronic control and viewing (eliminates flicker). The receiver beam must do likewise, sweeping every other line precisely in sequence with the camera. In the first pass, called a **field,** the beam must start in the upper left-hand corner and trace every odd line, ending at the middle of the bottom line for a total of 262½ lines (½ of 525). The beam must then return to the top center of the screen and sweep each even line in sequence, completing 262½ lines of the field at the right end of the bottom line.

In the process, the beam must excite the fluorescent screen with the correct intensity indicated by the corresponding portions of the video waveform. At the end of each line, the beam is blanked and must be rapidly returned to the left to start the next line of the picture. When the beam reaches the bottom of the screen, it must be blanked again and rapidly returned to the correct position (left or middle) at the top to sweep the next field. It must complete a **field** (262½ lines) in precisely 1/60 of a second and a full **frame** (complete picture) in 1/30 of a second.

Q15-54. There are _____ lines to a field and _____ fields to a frame.

Q15-55. The sweep of the receiver beam must be _____ with the _____ of the camera beam.

Q15-56. The start and the position of each scan line on the crt screen are controlled by the _____ _____.

Moving the Electron Beam

You know that a negative voltage repels and a positive voltage attracts electrons. The cathode-ray tube (**crt**) uses this effect to send an electron beam to the screen and control its movement.

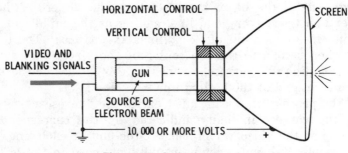

Fig. 15-27. Cathode-ray tube.

The basic construction and connections of a crt are illustrated in Fig. 15-27. At the left end is shown an electron gun which shoots a narrow stream of electrons toward the screen. To speed the electrons on their way, the inner surface of the flared portion of the tube has a conductive coating energized with a voltage that is several thousand volts positive with respect to the electron source.

The crt is connected to the output of the last video amplifier from which is received the video and blanking signals. The crt element which controls the number of electrons responds to the varying amplitude of the video and releases the quantity required. This element also stops the flow of electrons when the blanking pulse appears.

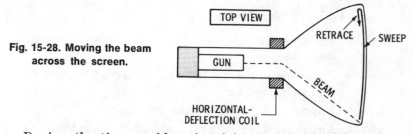

Fig. 15-28. Moving the beam
across the screen.

During the time a video signal is present, the beam must
be moved from left to right across the screen. When the beam
reaches the right side, the blanking pulse shuts off the electrons and the beam moves back to start the next line.

A **horizontal-deflection coil** wrapped around the neck of
the tube moves the beam from side to side. Current moving
through the coil sets up a magnetic field which has an attracting and a repelling effect on electrons similar to positive and
negative voltage. The stronger the field, the greater is its
effect on the beam. To increase the strength of the field
requires an increase in current through the coil. Fig. 15-28
shows the beam deflected to the left.

The change in strength of the magnetic field during a sweep
must coincide with the time duration of the scanned line. A
gradual rise of current within the coil during this time period
accomplishes this. The starting time is triggered by the sync
pulses that ride on the blanking pulses. If the current decreases rapidly at the end of each line (during the blanking
pulse), the sudden drop in magnetic field strength returns
the beam to the left very quickly.

A similar magnetic field is set up by a second coil (the
vertical-deflection coil) which controls the movement of the
beam line by line from the top of the screen to the bottom. On
completion of a field, the beam quickly retraces to the top.

Q15-57. The electron beam in the crt is generated by
a(an) _____ _____.

Q15-58. Electrons are drawn to the screen of the crt by
a(an) _____ _____.

Q15-59. _____ _____ move the crt beam.

Q15-60. An increase in current through a deflection coil
(increases, decreases) the magnetic field.

WARNING: A vacuum exists inside a cathode-ray tube, causing tremendous pressures to be exerted toward the center of the tube. If dropped, scratched, or carelessly jarred, the crt may implode (explode inward), scattering pieces of glass and metal at a tremendous velocity.

Sync Control Circuits

The beam movement is accomplished by steadily increasing the current flow in each of the deflection coils during precise time intervals. The starting times for these intervals are controlled by the sync pulses (Fig. 15-29).

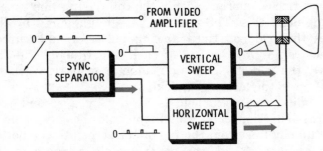

Fig. 15-29. Sync control circuits.

Sync Separator—Video waveforms arrive at the sync separator from the video amplifier. There is one narrow sync pulse for each line of scan. This pulse is intended to control the starting time of each horizontal sweep across the screen. When one field of 262½ lines has been completed, the video waveforms are followed by a sync pulse many times wider than the horizontal sync pulses.

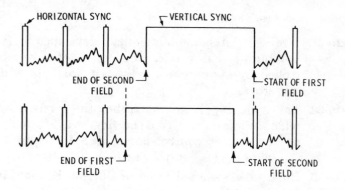

HORIZONTAL SYNC   VERTICAL SYNC

END OF SECOND
FIELD

START OF FIRST
FIELD

END OF FIRST
FIELD

START OF SECOND
FIELD

Fig. 15-30. Sync pulses.

This wide pulse is the trigger that develops a vertical sweep to move the beam from line to line down the face of the screen. Every other vertical sync pulse starts in the middle of a video waveform, accounting for 262½ lines in each field of interlace scanning.

Fig. 15-30 shows the comparative widths of the horizontal and vertical sync pulses and the relative starting times of the first and second fields. The sync pulses are removed from the complete video waveform by the sync separator.

The narrow and wide pulses are distributed to the appropriate sweep circuits (horizontal and vertical) after sync separation. This is accomplished by capacitor and resistor combinations which can distinguish between voltage waveforms with short time durations and those with long durations. The short sync pulses are sent to the **horizontal-sweep circuit** and the long pulses to the **vertical-sweep circuit**.

Q15-61. Timing of the magnetic field developed in the _____ and _____ deflection coils is controlled by _____ _____ extracted from the video waveform.

Q15-62. The sync separator separates the narrow and wide timing pulses from the video waveform. The narrow pulses control _____ deflection and the wide pulses control _____ deflection.

Q15-63. Horizontal-sync pulses occur during the _____ portion of the video waveform.

Sweep Circuits

The two sweep circuits (horizontal and vertical) generate a linear rising voltage each time they receive a sync pulse. The **horizontal-sweep** circuit is triggered 525 times during the same time the **vertical-sweep circuit** is triggered twice.

Horizontal-Sweep Circuit—Horizontal sweep is produced by an oscillator which generates a slowly rising and rapidly decaying sawtooth waveform, whether the set is tuned to a transmitting station or not. This accounts for the **raster** (lines on the screen) when the tv receiver is on but no signal is being received.

The purpose of the sync pulse is to trigger the oscillator so that oscillations start at the same time as the line scan in the camera. Capacitor and resistor combinations convert the oscillations to the sawtooth waveshapes shown in the diagram in Fig. 15-31. Rise time of the sawtooth causes the current in the horizontal-deflection coil to increase gradually, moving the beam across the screen in step with the line scan in the camera. At the end of the line, coil current decreases rapidly, returning the beam (which is now blanked) to the

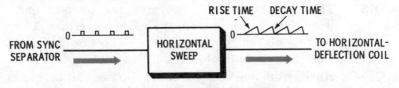

Fig. 15-31. Horizontal sweep circuit.

left side of the screen. There are 525 lines to each frame, so the frequency of the horizontal oscillator must be 15,750 hertz.

Vertical-Sweep Circuit—The vertical-sweep oscillator and amplifier are almost identical to those in the horizontal-sweep section. The main difference is that the frequency of oscillation is much lower—60 times a second, to match the frequency at which each field is swept. The rise time (plus a short decay time) of the vertical sawtooth lasts for 1/60 of a second before another vertical-sync pulse arrives to start the next waveform. Gradual increase in current in the vertical-deflection coil moves the beam from the top to the bottom of the screen. The decay of the vertical sawtooth waveform brings the beam back to the top in time for the next sync pulse.

Height Control—In most tv receivers a **height** control varies the setting of a variable resistor in the vertical-sweep stage. Adjustment of the control moves the starting position of the sweep up or down. The resistor controls the amount of initial current that flows through the coil.

High-Voltage Power Supply

The several thousand volts required for the crt are developed in the high-voltage power supply, a group of circuits usually contained in a metal cage inside the set. A diode discharging a capacitor through a transformer produces the high voltage for the crt. Even after the set has been turned off, the capacitor can retain its charge for some time. Precautions should, therefore, be taken if work must be done inside this protective cage.

WARNING: There can be up to 30,000 volts connected to a plug-in on the flared side of the crt. The voltage is applied through a heavily insulated conductor. Even though the set is turned off, approach the sides of the cathode-ray tube with extreme caution.

Q15-64. What is the horizontal sync-pulse frequency?

Q15-65. What is the frequency of the vertical-sync pulses?

Q15-66. What are two precautions you should observe when near a cathode-ray tube?

WHAT YOU HAVE LEARNED

1. Television transmitters and receivers, like any other electronic equipment, consist of circuits designed to accomplish specific functions. Although there are a large variety of circuits, they all operate in accordance with a basic concept—the effect that voltage, current, and electronic components have on each other. These basic effects can be used to analyze any circuit, providing the student understands the underlying principles of each.

2. A television transmitter consists of two sections. One section uses a camera to scan a scene, and a group of circuits to modulate a carrier frequency with the image. The other section takes the output from a microphone and uses it to modulate a second carrier frequency.

3. Sound is superimposed on its carrier by frequency modulation. The procedure is one in which the frequency of the carrier varies in accordance with the amplitude of the sound—decreasing during the negative portions of the audio cycle and increasing during the positive portions. The fm carrier is then amplified to the required power level and fed to the antenna.

4. Video is obtained from the camera as it scans a scene with an electron beam, one line at a time. The video signal, with the addition of blanking and synchronizing pulses, is amplified and then used to modulate the picture carrier frequency. The amplitude-modulated carrier is raised to a specified power level and then fed to the antenna.

5. Video and sound carriers are of a high frequency and, therefore, have short wavelengths. These travel through the atmosphere as either space or sky waves. Short space waves travel on a line-of-sight path and cannot be received beyond the horizon. Sky waves enter the ionosphere where their paths are bent by an amount depending on the angle of entry. If the entry angle is small, the sky wave returns to the earth and can be received.

6. A tv receiver contains many circuits that can be grouped into a few electronic functions. These include the rf section (front-end), if section, sound section, video section, vertical-sync control, horizontal-sync control, cathode-ray tube, and low- and high-voltage power supplies. Many of the functions are similar to those found in a radio receiver.

7. The rf amplifier, mixer, and oscillator select the desired channel among the many appearing on the antenna and convert the sound and video-carrier frequencies to appropriate intermediate frequencies.

8. The sound section, containing an if stage for amplification, a detector (or discriminator) for removal of the audio component, and audio amplifiers for further amplitude gain, processes the signal for operation of the speaker.

9. The video section contains similar circuits to extract the video signal and amplify it to a level required for operating the beam-control portion of the crt.

10. Vertical- and horizontal-sync pulses are taken from the video signal and channeled through corresponding vertical- and horizontal-sweep circuits. The sawtooth waveforms developed by these circuits control the movement of the electron beam, causing the image to be placed on the screen, a line at a time, in precise synchronization with the camera scan beam.

11. Highly dangerous voltages exist on the cathode-ray tube and inside the protective cage of the high-voltage power supply. Extreme caution should be used when working in these areas.

12. Most of all, you have learned a great deal more about electronics. You should now have acquired a fairly good

understanding of how electronic circuits work. A mental image of circuit and equipment operation will give you a solid reference against which you can base the details of current, voltage, resistance, inductance, and capacitance principles that are explained in following volumes. You need a clear understanding of these principles if you plan to become technically competent when working with electronic equipment.

Index